Lecture Notes in Mathematics 1592

Editors:
A. Dold, Heidelberg
B. Eckmann, Zürich
F. Takens, Groningen

Karl Wilhelm Breitung

Asymptotic Approximations for Probability Integrals

Springer-Verlag
Berlin Heidelberg New York
London Paris Tokyo
Hong Kong Barcelona
Budapest

Author

Karl Wilhelm Breitung
Department of Civil Engineering
University of Calgary
2500 University Drive, N.W.
Calgary, Alberta T2N 1N4, Canada

Mathematics Subject Classification (1991): 41A60, 41A63, 60F10, 60G15, 60G70, 62N05, 62P99, 90B25

ISBN 3-540-58617-2 Springer-Verlag Berlin Heidelberg New York

CIP-Data applied for

Typesetting: Camera ready by author
SPIN: 10130174 46/3140-543210 - Printed on acid-free paper

Preface

In this lecture note asymptotic approximation methods for multivariate integrals, especially probability integrals, are developed. It is a revised version of my ***habilitationsschrift*** "Asymptotische Approximationen für Wahrscheinlichkeitsintegrale". The main part of this research was done when I worked at the Technical University and the University of Munich.

The motivation to study these problems comes from my work in the research project "Zuverlässigkeitstheorie der Bauwerke" (reliability theory of structures) at the Technical University of Munich in the department of civil engineering. For the tolerant support of the mathematical research in this project I would like to thank Prof. Dr.-Ing. H. Kupfer.

I am grateful to Prof. Dr.-Ing. R. Rackwitz and Prof. Dr.-Ing. G.I. Schuëller that they made me clear the engineering topics of reliability theory and helped me in my research. Further I would like to thank my former colleagues at the Technical University of Munich and at the University of Munich which supported me during my work at these universities: Dr.-Ing. B. Fießler, Dr. M. Hohenbichler, Dr. A. Rösch, Dr. H. Schmidbauer and Dr. C. Schneider. Additionally I would like to express my gratitude to Prof. Dr. F. Ferschl for pointing out occasional errors and misprints in the original German version.

The major part of this revision was made when I stayed as visiting fellow at the University of New South Wales in 1991. I would like to thank especially Prof. A. M. Hasofer for his help and for his kind invitation to the University of New South Wales and to express my delight at having worked there.

For their help and discussions I thank wholeheartedly Prof. Dr. F. Casciati, Prof. Dr. L. Faravelli (both University of Pavia), Prof. Dr. P. Filip (FH Bochum), Prof. Dr. K. Marti (University of the Federal Armed Forces at Neubiberg) and Prof. Dr. W.-D. Richter (University of Rostock). Prof. Dr. M. Maes (University of Calgary) knows what I mean.

For eliminating the worst bugs in my English I thank Poul Smyth, the Irish poet of the Bermuda triangle, and for making a cover design Ringo Praetorius, the executioner of Schichtl at the Oktoberfest. Unfortunately the publisher and the series editors decided not to use this cover design.

Finally a short comment about the mathematical level of this note should be made. It is intended also for mathematically interested reliability engineers. Probably, therefore, the mathematicians will complain about the low level and the inclusion of too much elementary material and the engineers will go the other way.

Calgary, September 1994

Karl Wilhelm Breitung

Contents

Notation

The set of the natural numbers is denoted by $\mathbb{N}$ and the set of complex numbers by $\mathbb{C}$. The n-dimensional euclidean space is denoted by $\mathbb{R}^n$. For the set of the vectors in $\mathbb{R}^n$ with all components being positive we write $\mathbb{R}^n_+$. A vector in $\mathbb{R}^n$ is written as $\boldsymbol{x}$ and the zero vector $(0, \ldots, 0)$ as $\mathbf{o}$. The transpose of $\boldsymbol{x}$ is written as $\boldsymbol{x}^T$. The unit vector in direction of the x_i-axis is denoted by $\boldsymbol{e}_i$. For the euclidean norm of a vector $\boldsymbol{x}$ we write $|\boldsymbol{x}|$ and for the scalar product of two vectors $\boldsymbol{x}$ and $\boldsymbol{y}$ we use $\langle \boldsymbol{x}, \boldsymbol{y} \rangle$. The subspace of $\mathbb{R}^n$ spanned by k vectors $\boldsymbol{a}_1, \ldots, \boldsymbol{a}_k$ is written as $\text{span}[\boldsymbol{a}_1, \ldots, \boldsymbol{a}_k]$. For the orthogonal complement of a subspace $U \subset \mathbb{R}^n$ we write $U^\perp$.

An $n \times m$-matrix is written with bold Roman letters: $\boldsymbol{A}$,$\boldsymbol{B}$,... The n-dimensional unity matrix is denoted by $\boldsymbol{I}_n$ and an $n \times k$ matrix consisting of zeros by $\mathbf{o}_{n,k}$. The cofactor matrix $\boldsymbol{C}$ of an $n \times n$ matrix $\boldsymbol{A}$ is the $n \times n$ matrix $\boldsymbol{C} = ((-1)^{i+j} \det(\boldsymbol{A}_{ij}))_{i,j=1,\ldots,n}$ with $\boldsymbol{A}_{ij}$ being the $(n-1) \times (n-1)$ matrix obtained from $\boldsymbol{A}$ by deleting the i-th row and the j-th column. The rank of a matrix $\boldsymbol{B}$, i.e. the number of its linearly independent column vectors, is denoted by $\text{rk}(\boldsymbol{B})$.

The probability of an event A is denoted by $\mathbb{P}(A)$. An one-dimensional random variable is denoted by a capital Roman letter: $X, Y, \ldots$ and for n-dimensional random vectors bold capital Roman letters are used: $\boldsymbol{X}, \boldsymbol{Y}, \ldots$. For the probability density function and the cumulative distribution function of a random variable we write p.d.f. and c.d.f. respectively. The expected value of a random variable X is written as $\mathbb{E}(X)$ and its variance as $\text{var}(X)$. The covariance between X and Y is denoted by $\text{cov}(X, Y)$.

A function $f : D \to \mathbb{R}$ on an open set $D \subset \mathbb{R}^n$ is called a C^1-function if all partial derivatives of first order exist and are continuous. Analogously by induction C^r-functions $(r > 1)$ are defined. A function $f : D \to \mathbb{R}$ on an open set $D \subset \mathbb{R}^n$ is called a C^r-function if all partial derivatives of order $r-1$ exist and are continuously differentiable. Further a function $f : D \to \mathbb{R}$ on a closed set $D \subset \mathbb{R}^n$ is called a C^r-function if there is an open set $U \subset \mathbb{R}^n$ with $D \subset U$ such that f is defined on U and f is according to the definition above a C^r-function.

A function $T : \mathbb{R}^n \to \mathbb{R}^m, \boldsymbol{x} \mapsto T(\boldsymbol{x}) = (t_1(\boldsymbol{x}), \ldots, t_m(\boldsymbol{x}))$ is called a C^r-vector function if all component functions $t_1(\boldsymbol{x}), \ldots, t_m(\boldsymbol{x})$ are C^r-functions. For a function $f : \mathbb{R}^n \to \mathbb{R}$ the first partial derivatives with respect to the variables x_i $(i = 1, \ldots, n)$ at the point $\boldsymbol{x}$ are denoted by $f^i(\boldsymbol{x})$ or $\frac{\partial f(\boldsymbol{x})}{\partial x_i}$ and the gradient of by $\nabla f(\boldsymbol{x})$. The second derivatives of this function with respect to the variables x_i and x_j $(i, j = 1, \ldots, n)$ at the point $\boldsymbol{x}$ are denoted by $f^{ij}(\boldsymbol{x})$ or $\dfrac{\partial^2 f(\boldsymbol{x})}{\partial x_i \partial x_j}$ and its Hessian by $\boldsymbol{H}_f(\boldsymbol{x})$.

For functions of the form $f(\boldsymbol{x}, \boldsymbol{y})$ by $\nabla_{\boldsymbol{x}} f(\boldsymbol{x}, \boldsymbol{y})$ the gradient with respect to the vector $\boldsymbol{x}$ is denoted. In the same way $\nabla_{\boldsymbol{y}} f(\boldsymbol{x}, \boldsymbol{y})$ means the gradient with respect to the second vector $\boldsymbol{y}$. The divergence of a C^1-vector function $\boldsymbol{u}(\boldsymbol{x})$ is denoted by $\text{div}(\boldsymbol{u}(\boldsymbol{x}))$.

Chapter 1

Introduction

1.1 The Evaluation of Multivariate Integrals

In many fields of applied mathematics it is necessary to evaluate multivariate integrals. In a few cases analytic solutions can be obtained, but normally approximation methods are needed. The standard methods here are numerical and Monte Carlo integration. In spite of the fact that computation time today is accessible plentiful, such procedures are sufficient for many problems, but in a number of cases they do not produce satisfactory results. Such integrals appear for example in structural reliability, stochastic optimization, mathematical statistics, theoretical physics and information theory.

If we consider an integral in the form

$$\int_F f(x)\,dx \tag{1.1}$$

with $F \subset \mathbb{R}^n$ and $f : F \to \mathbb{R}$, there are three main causes, which might make numerical or Monte Carlo integration difficult:

1. The dimension n of the integration domain is large.

2. The domain F has a complicated shape.

3. The variation of f in the domain F is large.

Often not only the integral itself, but its derivatives with respect to parameters are of interest. The most general case is that the integrand $f(x)$ as well as the integration domain F depend on parameters.

Another important point is that often not only one integral has to be computed, but the behavior of the integral under parameter changes is of interest. Then many evaluations of the integral are necessary. A similar problem occurs if some sort of optimization should be made. Further in reliability problems with random finite elements (see for example [4]) even now the necessary computing time can be prohibitive.

Methods for analytic approximations have been developed in different fields and sometimes due to the specialization nowadays in science some of them have been rediscovered at least once or twice. Therefore it is attempted here to give a list of the available textbooks in this field and certainly for the field of multivariate Laplace methods an overview of the relevant results.

The basic idea of such methods is that instead of integrating over the whole domain F, points or subsets are identified from whose neighborhoods the main contributions to the integral come from. Therefore instead of integrating the numerical problem is then to find these sets by some optimization method.

Such approximation methods are not a solution for all problems. Their efficiency and usefulness depends on the problem. If the underlying assumptions about the structure of the integral are not fulfilled to some degree, the use of other schemes might be better. In some cases then the use of asymptotic approximation methods, which are described in this book, perhaps in combination with the aforementioned methods, is advisable.

1.2 Structural Reliability

One field, in which such concepts had been very successful, is structural reliability. In the following we will give a short outline of this field.

The first proposals to use probabilistic models for structural reliability are from [98], [78] and [64], but not before the sixties such problems were studied more intensively. Then in the last thirty years in mechanical and civil engineering probabilistic methods were developed for the calculation of the reliability of components and structures, since pure deterministic methods were not adequate for a number of problems. Textbooks about such methods are [10], [16], [17], [39], [84], [89], [99], [124] and [128].

In structural reliability the computation of multivariate integrals is an important problem. The studies in this lecture note were motivated by the problems in this field, since standard methods did not lead to satisfactory results.

At the beginning the random influences acting on a structure were modelled simply by two time-invariant random variables, a load variable L and a resistance variable R. If $L \geq R$, the structure failed, if $L < R$, it remained intact.

But soon it was clear that such a model was far too simple even for components of a structure. Even if the time influence is neglected, for a sufficient description of the random influences a random vector $\boldsymbol{X} = (X_1, \ldots, X_n)$ with a large number n of components is needed. In general one part of this vector is composed of load variables and the other of resistance variables.

If now the p.d.f. $f(\boldsymbol{x})$ of the random vector $\boldsymbol{X}$ which models the random influences on a structure is known and the conditions for failure can be expressed as a function of the vector, the probability for failure can be calculated. Then the integration domain is given by a function $g(\boldsymbol{x})$ in the form $\{\boldsymbol{x}; g(\boldsymbol{x}) \leq 0\}$. The function $g(\boldsymbol{x})$ describes the state of the structural system under consideration. If $g(\boldsymbol{x}) > 0$, the system is intact and if $g(\boldsymbol{x}) \leq 0$, the system fails.

The failure domain is denoted by $F = \{\boldsymbol{x}; g(\boldsymbol{x}) \leq 0\}$, the safe domain by $S = \{\boldsymbol{x}; g(\boldsymbol{x}) > 0\}$ and the limit state surface, the boundary of F by $G = \{\boldsymbol{x}; g(\boldsymbol{x}) = 0\}$.

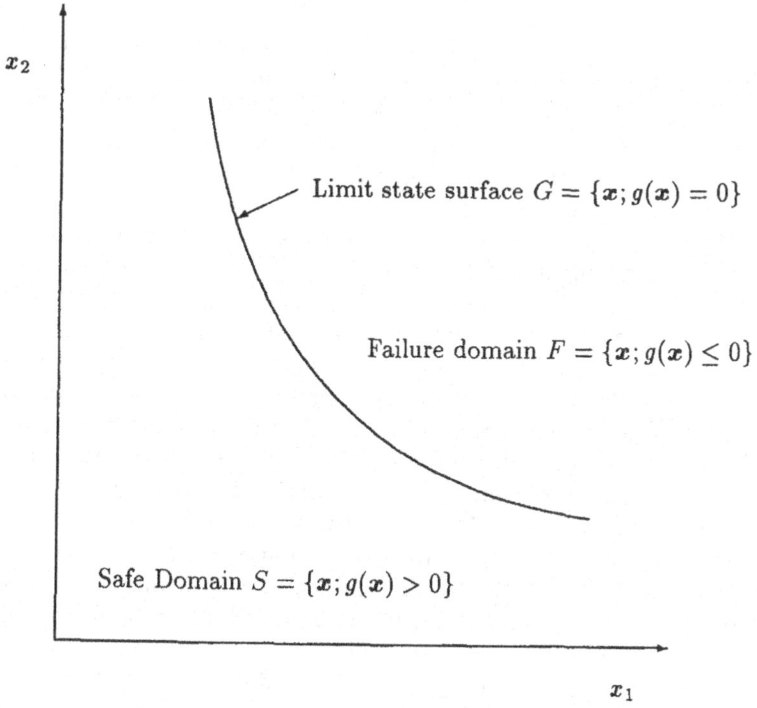

Figure 1.1: Failure domain, limit state surface and safe domain

Then the problem is to compute the probability of failure

$$\mathbb{P}(F) = \int_{g(\boldsymbol{x}) \leq 0} f(\boldsymbol{x})\, d\boldsymbol{x}. \tag{1.2}$$

Often we can distinguish two types of random variables in the random vector $\boldsymbol{X}$. Firstly random variables, which can not be changed by the design of the structure, such as loads acting on it; secondly variables, which can be influenced by the design, for example the strength of concrete.

It was tried first to use the standard methods for the calculation of such integrals mentioned above to compute these multivariate probability integrals. In numerical or Monte Carlo integration then the integral is approximated in the form

$$\mathbb{P}(F) \approx \sum_{i=1}^{N} w(\boldsymbol{x}_i) f(\boldsymbol{x}_i). \tag{1.3}$$

Here the $\boldsymbol{x}_i$'s are points in F, determined by a deterministic or random mechanism and the $w(\boldsymbol{x}_i)$ are their respective weights. The main difficulties in computing this integral are that in general the probability density $f(\boldsymbol{x})$ is small in the failure domain and that the shape of the domain is not explicitly known, but given implicitly by the function $g(\boldsymbol{x})$; therefore standard techniques have difficulties in finding a scheme for creating enough points in F. Soon it became clear that in their usual form they were not suited to this problem, since they were too time consuming and inaccurate.

Certainly now with the increasing capacity of computers it is possible to solve more and more reliability problems, without spending any effort on refining methods. But the author thinks that it is absolutely wrong to say that we just have to wait until the computing capacity is large enough to make it possible to solve more complex problems. Since the main point in structural reliability and other fields is not only the computation of probabilities, but the gain of insight in the structure of the problem. If an approximate analytic solution is found, the important influences on the failure mechanism become often much clearer.

The usual formulation of the problem, compute $\mathbb{P}(F)$, given the density $f(\boldsymbol{x})$ and the limit state function $g(\boldsymbol{x})$, is misleading. In reality the probability distribution is known only approximately and also the limit state function is only an appproximation. The problem of estimating such a function by experimental design is considered in [58] and [32]. Der Kiureghian [44] and Maes [90] treat the problem of structural reliability under model uncertainties.

Furthermore in a problem of structural reliability the probability of failure depends on a number of design parameters in the structure. Some of these parameters can be changed by altering the design of the structure, others not. Therefore the real problem here is more complex than it appears at the first glance.

Therefore a more realistic probabilistic model is in the form

$$\mathbb{P}(F|\boldsymbol{\vartheta}) = \int\limits_{g(\boldsymbol{x}|\boldsymbol{\vartheta})\leq 0} f(\boldsymbol{x}|\boldsymbol{\vartheta})\, d\boldsymbol{x}. \tag{1.4}$$

Here all functions depend on a parameter vector $\boldsymbol{\vartheta} = (\vartheta_1, \dots, \vartheta_k)$.

1. $g(\boldsymbol{x}|\boldsymbol{\vartheta})$ is the limit state function for the parameter value $\boldsymbol{\vartheta}$.

2. $f(\boldsymbol{x}|\boldsymbol{\vartheta})$ is the probability density function of the random vector $\boldsymbol{X}$ for the parameter value $\boldsymbol{\vartheta}$.

The usual formulation is that the parameter vector $\boldsymbol{\vartheta}$ is assumed to be fixed and known. Then for this fixed vector $\boldsymbol{\vartheta} = \boldsymbol{\vartheta}_0$ the probability $\mathbb{P}(F|\boldsymbol{\vartheta}_0)$ is computed. But in the more general setting this is only one of the interesting quantities. If the values of the parameters are known and can be influenced by the design of the structure, one problem can be to optimize the structure in some sense. In reliability the quantity to be minimized is usually the probability of failure, but there are restrictions on the possible parameter values.

The following quantities and distributions may be of interest:

1. The partial derivatives of $\mathbb{P}(F|\boldsymbol{\vartheta})$ with respect to the components $\vartheta_1, \ldots, \vartheta_k$ of the parameter vector.

2. The distribution of $\mathbb{P}(F|\boldsymbol{\vartheta})$ if there is a probability distribution of $\boldsymbol{\vartheta}$.

3. The asymptotic form of the conditional distribution of $\boldsymbol{X}$ under the condition $g(\boldsymbol{X}|\boldsymbol{\vartheta}) < 0$.

Until now only the time-invariant reliability problem is solved sufficiently. Often time-variant problems are transformed into a time-invariant form, for example by modelling a random process only by its extreme value distribution. This is related to the fact that the theory of multivariate stochastic processes and their extreme value theory is restricted mainly to Gaussian processes.

In probabilistic models for structural reliability we have two different probability theory and statistical inference. With methods of probability theory and other mathematical concepts the structure of the mathematical model is studied. The other problem is the relation of the model to reality. A common feature of the majority of books and papers in structural reliability is that the statistical problems are often neglected.

We will discuss shortly the problems of statistical inference in reliability modelling. Since in reliability problems data are often sparse or do not exist at all, conclusions are in many cases based on a lot of assumptions, whose correctness can not be proved. The results, which are obtained, are difficult to check. For example if scientist A computes a failure probability of 10^{-5} for a single building during one year and scientist B instead a probability of 10^{-6}, who is right? An approach for a solution might be the development of probabilistic models, which make more specific predictions (see for example [30] and [34]). This would mean that not only failure probabilities are computed, but the probability distributions of other events, which are connected with the occurrence of failures, but can be observed more frequently. By such an approach, we get an iterative process of model building, prediction, observation and model improvement.

In the book of Matheron [97] the general problem of probabilistic models in science is discussed. The importance of this book is that here it is made clear that the justification and the acceptance of these methods in science comes from the fact that they give for many problems satisfactory solutions, but not that they are correct descriptions of the reality. This might be quite unintelligible for statisticians who have never been involved in any applied work as they have lived in a pure academic environment. But anyone who works at problems which have some connection with the real world outside Academia will probably agree that this is not so wrong at all.

An alternative method in coping with uncertainty in structural reliability is for example fuzzy set theory (see [6]). In fuzzy set methods uncertainties, which are not of a probabilistic nature, can be modelled. A drawback is that there is no clear rule how to incorporate additional information as it is done for example in the Bayesian concept by Bayes' theorem.

A further alternative concept is convex modelling (see [9]). In convex modelling no specific probability distribution for the random influences is assumed and only bounds for admissible influences are derived.

Another problem of statistical inference in reliability is that the usual statistical estimation methods are focused on fitting distributions to the central part of the data and not to the tails. Here modified estimation procedures should be used (see [40]), which give more weight to the fit in the distribution tails, since in reliability calculation the risk is mainly in underestimating the extremes of random influences and not in making wrong estimates about their means.

1.3 Stochastic Optimization

A mathematical field, where similar problems occur, is stochastic optimization. Here a given stochastic system, for example a network, a technical structure or a queuing system, should be optimized in its performance, which is described by a function of the design parameters, but is in general not known analytically. Basic concepts of stochastic optimization can be found in the book of Rubinstein [123].

For a given stochastic system we have an integral in the form

$$I(\boldsymbol{\vartheta}) = \int_{\mathbb{R}^n} R(\boldsymbol{x}|\boldsymbol{\vartheta}) f(\boldsymbol{x}|\boldsymbol{\vartheta})\, d\boldsymbol{x}, \tag{1.5}$$

and a set of k constraint functions

$$g_1(\boldsymbol{\vartheta}) = \ldots = g_k(\boldsymbol{\vartheta}) = 0. \tag{1.6}$$

Here $\boldsymbol{\vartheta}$ is an m-dimensional parameter vector, who can achieve values in a subset $V \subset \mathbb{R}^m$. $f(\boldsymbol{x}|\boldsymbol{\vartheta})$ with $\boldsymbol{x} \in \mathbb{R}^n$ is a p.d.f., $R(\boldsymbol{x}|\boldsymbol{\vartheta})$ is a function, which describes the performance of the system for the given values $\boldsymbol{\vartheta}$ and $\boldsymbol{x}$. The functions $g_1(\boldsymbol{\vartheta}), \ldots, g_k(\boldsymbol{\vartheta})$ give restrictions for the possible values of $\boldsymbol{\vartheta}$ and describe usually costs or available resources. Then $I(\boldsymbol{\vartheta})$ is the performance of the system for the parameter vector $\boldsymbol{\vartheta}$. In stochastic optimization for fixed values of the parameter $\boldsymbol{\vartheta}$ the values of at least some of the functions are not known exactly and have to be estimated by some procedure. Usually here the problem is considered that $g_i(\boldsymbol{\vartheta}) = \mathbb{E}(h_i(\boldsymbol{X}))$ and only random samples are available from which the expected value $\mathbb{E}(h_i(\boldsymbol{X}))$ has to be estimated.

In this formulation the task is to minimze the integral $I(\boldsymbol{\vartheta})$ under the restrictions in equation (1.6). For this purpose it is necessary to evaluate the function $I(\boldsymbol{\vartheta})$ and at least some of its sensitivities (derivatives, gradients, Hessians, etc.)

with respect to changes in the parameter vector ϑ. To compute these, various methods can be used. Here often Monte Carlo methods are adequate.

Applications of stochastic optimization methods in structural design are given in [96] and [95].

1.4 Large Deviations and Extreme Values

Large deviation theory studies the asymptotic behavior of the probabilities $\mathbb{P}(\lambda A)$ as $\lambda \to \infty$; here $\mathbb{P}(.)$ is a probability measure on a measurable space $(E, \mathcal{B})$ and A is an subset of E. If for example the underlying space is the n-dimensional Euclidean space $\mathbb{R}^n$, then such probabilities are given by n-dimensional integrals. The case of the standard normal probability measure is considered in the papers of Richter ([114], [14], [115], [116], [118] and [107]).

Similar questions appear if we study the asymptotic distribution of the sum $\sum_{i=1}^n X_i$ of random variables with mean zero (see [18], [19] and [119]).

Extreme value theory is concerned with the maxima (or minima) of sequences of random variables or of a stochastic process. If we consider a sequence $X_1, X_2, \ldots$ of i.i.d random variables, the classical question of extreme value theory was to find under which conditions by a suitable scaling and shifting a non-degenerate limit distributions exist for the sequences

$$Y_n = \max(X_1, \ldots, X_n) \text{ resp. } Z_n = \min(X_1, \ldots, X_n). \tag{1.7}$$

The classical textbook in this field is the book [68] of Gumbel written 1958. The textbook of Leadbetter, Lindgren and Rootzén [82] gives in the first part an overview of further development of the extreme value theory for sequences.

In the case of a stochastic process $X(t)$ we consider the random variables

$$Y(T) = \max_{0 \le t \le T} X(t) \quad \text{resp. } Z(T) = \min_{0 \le t \le T} X(t). \tag{1.8}$$

Here similar results as for sequences can be obtained, see [43] and [82]. The results in these books are usually derived by approximating the process by a suitably chosen sequence of random variables. An alternative method, based on sojourn times, is outlined in [12].

1.5 Mathematical Statistics

Similar questions arise in mathematical statistics. Here asymptotic methods play an important role in investigating large sample behavior. An overview of such applications can be found in the book of Barndorff-Nielsen and Cox [7]. Here a sequence of i.i.d. random variables $X_1, X_2, \ldots$ is given and for functions $f(X_1, \ldots, X_n)$ as $n \to \infty$ asymptotic approximations are sought. A standard example are the asymptotic behaviors of the sample mean $\bar{X} = n^{-1} \sum_{i=1}^n X_i$ and the sample variance $\bar{S} = (n-1)^{-1} \sum_{i=1}^n (X_i - \bar{X})^2$, which converge under some regularity conditions to the respective moments of the X_i's.

If the random variable X_1 has the p.d.f. $f(x)$ the joint p.d.f. of the first n random variables is $\prod_{i=1}^n f(x_i)$ and the log likelihood function is $\ln(\prod_{i=1}^n f(x_i)) = \sum_{i=1}^n \ln(f(x_i))$. The asymptotic behavior of these function can be studied using the Laplace method. In the case of random vectors $\boldsymbol{X}_1, \boldsymbol{X}_2, \ldots$ we need again results about the asymptotic structure of multivariate integrals.

Additional importance have asymptotic approximation methods in Bayesian statistics, where the derivation of the posterior distribution requires often the evaluation of multivariate integrals. The first who used the Laplace method outlined in this book for such problems, i.e. the derivation of posterior distributions, was Lindley [86]. Further results are given in [131].

1.6 Contents of this Lecture Note

In chapter 2 first some results from linear algebra and analysis are given for reference and some new results about sufficient conditions for constrained extrema and multivariate parameter dependent integrals are derived. In the third chapter a short outline of the basic concepts of asymptotic analysis is given. Then in the fourth univariate and in the fifth multivariate Laplace type integrals are studied. In the sixth these results are applied to normal random vectors. In the seventh chapter it is shown that such approximations can be made also for non-normal random vectors. In the last chapter then asymptotic approximations for crossing rates of differentiable stationary Gaussian vector processes are derived.

Chapter 2

Mathematical Preliminaries

2.1 Results from Linear Algebra

In this section some results from linear algebra, which are needed in the following and are not always in standard textbooks, are collected.

Let $U \subset \mathbb{R}^n$ be a linear subspace of $\mathbb{R}^n$ and $U^\perp$ its orthogonal complement. For each vector $\boldsymbol{x} \in \mathbb{R}^n$ there is a unique orthogonal projection onto this subspace. This means there exists a vector $\boldsymbol{u} \in U$ and a vector $\boldsymbol{w} \in U^\perp$ such that $\boldsymbol{x} = \boldsymbol{u} + \boldsymbol{w}$. This projection vector is obtained by multiplying $\boldsymbol{x}$ by the $n \times n$ projection matrix $\boldsymbol{P}_U$ of the subspace U.

Lemma 1 *Let U be a k-dimensional subspace of $\mathbb{R}^n$ and $U^\perp$ its orthogonal complement, an $(n-k)$-dimensional subspace. Then for the projection matrix $\boldsymbol{P}_U$ the following relations hold*

1. $\boldsymbol{P}_U = \boldsymbol{P}_U^2$, *i.e.* $\boldsymbol{P}_U$ *is idempotent,*
2. $\boldsymbol{P}_{U^\perp} = \boldsymbol{I}_n - \boldsymbol{P}_U$,
3. $\boldsymbol{P}_U^T = \boldsymbol{P}_U$.

PROOF: Follows from the definitions. □

Lemma 2 *Let $\boldsymbol{x}$ be an n-dimensional vector and $\boldsymbol{a}_1, \ldots, \boldsymbol{a}_k \in \mathbb{R}^n$ with $k \leq n$ be linearly independent vectors which are the columns of the $n \times k$ matrix $\boldsymbol{A} = (\boldsymbol{a}_1, \ldots, \boldsymbol{a}_k)$. Then we have for the unique orthogonal projection $\boldsymbol{P}_U \boldsymbol{x} = \sum_{i=1}^k \gamma_i \boldsymbol{a}_i$ of a vector $\boldsymbol{x}$ onto the subspace $\boldsymbol{U} = \mathrm{span}[\boldsymbol{a}_1, \ldots, \boldsymbol{a}_k]$ that*

1. $\boldsymbol{P}_U = \boldsymbol{A}(\boldsymbol{A}^T \boldsymbol{A})^{-1} \boldsymbol{A}^T$,
2. $\gamma^T = (\gamma_1, \ldots, \gamma_k)^T = (\boldsymbol{A}^T \cdot \boldsymbol{A})^{-1} \boldsymbol{A}^T \boldsymbol{x}$,
3. $\boldsymbol{P}_U \boldsymbol{x} = \sum_{i=1}^k \gamma_i \boldsymbol{a}_i = \boldsymbol{A} \cdot (\boldsymbol{A}^T \cdot \boldsymbol{A})^{-1} \boldsymbol{A}^T \boldsymbol{x}$.

PROOF: We have that $\boldsymbol{P}_U\boldsymbol{x} \in \text{span}[\boldsymbol{a}_1, \ldots, \boldsymbol{a}_k]$ due to definition. If we define the projection matrix as in 1., we get

$$\begin{aligned}(\boldsymbol{x} - \boldsymbol{P}_U\boldsymbol{x})^T\boldsymbol{P}_U\boldsymbol{x} &= \boldsymbol{x}^T\boldsymbol{P}_U\boldsymbol{x} - \boldsymbol{x}^T\boldsymbol{P}_U^T\boldsymbol{P}_U\boldsymbol{x} \\ &= \boldsymbol{x}^T\boldsymbol{P}_U\boldsymbol{x} - \boldsymbol{x}^T\boldsymbol{P}_U\boldsymbol{x} = 0.\end{aligned} \tag{2.1}$$

Therefore $\boldsymbol{P}_U\boldsymbol{x}$ is orthogonal to $\boldsymbol{x} - \boldsymbol{P}_U\boldsymbol{x}$ and the projection matrix is in fact given by $\boldsymbol{A}(\boldsymbol{A}^T\boldsymbol{A})^{-1}\boldsymbol{A}^T$. □

Lemma 3 *Let $\boldsymbol{a}_1, \ldots, \boldsymbol{a}_k \in \mathbb{R}^n$ be k linearly independent vectors with $k \leq n$ which are the columns of the $n \times k$ matrix $\boldsymbol{A} = (\boldsymbol{a}_1, \ldots, \boldsymbol{a}_k)$. Then the volume $\text{V}(P)$ of the k-dimensional parallelepiped $P = \{\boldsymbol{y}; \boldsymbol{y} = \sum_{i=1}^{k} \alpha_i\boldsymbol{a}_i$ with $0 \leq \alpha_i \leq 1$ for $i = 1, \ldots, k\}$ is given by*

$$\text{V}(P) = \sqrt{\det(\boldsymbol{A}^T \cdot \boldsymbol{A})} \; . \tag{2.2}$$

Here $\det(\boldsymbol{A}^T \cdot \boldsymbol{A})$ is the Gramian of the vectors $\boldsymbol{a}_1, \ldots, \boldsymbol{a}_k$.

PROOF: This result is proved by induction over n (see for example [103], p. 81 or [102], p. 181). □

Definition 1 *Let $\boldsymbol{A}$ be an $m \times n$ matrix. An $n \times m$-matrix $\boldsymbol{A}^-$ with the following properties*

1. *$\boldsymbol{A}\boldsymbol{A}^-$ is symmetric,*
2. *$\boldsymbol{A}^-\boldsymbol{A}$ is symmetric,*
3. *$\boldsymbol{A}\boldsymbol{A}^-\boldsymbol{A} = \boldsymbol{A}$,*
4. *$\boldsymbol{A}^-\boldsymbol{A}\boldsymbol{A}^- = \boldsymbol{A}^-$,*

is called a generalized inverse of $\boldsymbol{A}$.

If $\boldsymbol{A}$ has an inverse, then $\boldsymbol{A}^{-1} = \boldsymbol{A}^-$. Sometimes this generalized inverse is called Moore-Penrose-inverse. For every matrix $\boldsymbol{A}$ exists a unique generalized inverse (see [67], p. 24-25). If the linear equation system $\boldsymbol{A}\boldsymbol{x} = \boldsymbol{b}$ can be solved, then $\boldsymbol{A}^-\boldsymbol{b}$ is a solution of the system. A detailed discussion of generalized inverses and their properties can be found for example in [113]. For two special cases we have the following results.

Lemma 4 *Let $\boldsymbol{D}$ be an $n \times n$-diagonal matrix with diagonal elements d_{ii} ($i = 1, \ldots, n$). Then the generalized inverse $\boldsymbol{D}^-$ of $\boldsymbol{D}$ is an $n \times n$ diagonal matrix with the i-th diagonal element equal to d_{ii}^{-1} if $d_{ii} \neq 0$ and zero if $d_{ii} = 0$.*

PROOF: See [67], p. 28. □

Lemma 5 *Let $A = (a_1, \ldots, a_k)$ be an $n \times k$-matrix ($k \leq n$), such that for the k vectors a_i always $a_i^T \cdot a_j = \delta_{ij}$. Then the generalized inverse of A is given by*

$$A^- = A^T. \tag{2.3}$$

PROOF: From the definitions follows that AA^T and A^TA are symmetric. Using the definition of A we have

$$A^TA = (\delta_{ij})_{i,j=1,\ldots,k} = I_k. \tag{2.4}$$

From this follows that

$$AA^TA = AI_k = A, \tag{2.5}$$

and

$$A^TAA^T = I_kA^T = A^T. \tag{2.6}$$

□

The following lemma shows that under some conditions two quadratic forms associated with two symmetric matrices can be brought together in a diagonal form.

Lemma 6 *Let be A a symmetric positive definite $n \times n$ matrix and B a symmetric positive semidefinite $n \times n$ matrix. Then there exists a non-singular $n \times n$ matrix T and non-negative numbers $\mu_1, \ldots, \mu_n$ with $\det(T) = \det(A)^{-1/2}$ such that*

$$(xT)^T ATx = x^T x = \sum_{i=1}^{n} x_i^2, \tag{2.7}$$

$$(xT)^T BTx = \sum_{i=1}^{n} \mu_i x_i^2. \tag{2.8}$$

PROOF: See [112], p. 29. □

2.2 Results from Analysis

To measure the sensitivity of a function with respect to changes in the arguments usually the partial derivatives are calculated. But they depend on the scale used for the various variables and are not invariant with respect to linear transformations. A better measure for the sensitivity of a function with respect to changes in the variables are its partial elasticities.

Definition 2 *For a function $f : \mathbb{R}^n \to \mathbb{R}$ the partial elasticity $\epsilon_i(f(x))$ with respect to x_i at a point x with $f(x) \neq 0$ is defined by*

$$\epsilon_i(f(x)) = x_i \cdot \frac{f^i(x)}{f(x)}. \tag{2.9}$$

The partial elasticity gives approximately the change of the function value in percents if the value x_i of the i-th component is changed by 1%.

Let in the $\mathbb{R}^n$ a coordinate change by a rotation is made, i.e. the new unit vectors are taken as $\boldsymbol{a}_1, \ldots, \boldsymbol{a}_n$, where $\langle \boldsymbol{a}_i, \boldsymbol{a}_j \rangle = \delta_{ij}$ and $\boldsymbol{A} = (\boldsymbol{a}_1, \ldots, \boldsymbol{a}_n)$, For a vector $\boldsymbol{x} = (x_1, \ldots, x_n) = (\langle \boldsymbol{x}, \boldsymbol{e}_1 \rangle, \ldots, \langle \boldsymbol{x}, \boldsymbol{e}_n \rangle)$, then the new coordinates $\boldsymbol{y} = (y_1, \ldots, y_n) = (\langle \boldsymbol{x}, \boldsymbol{a}_1 \rangle, \ldots, \langle \boldsymbol{x}, \boldsymbol{a}_n \rangle)$ are given by $\boldsymbol{y} = \boldsymbol{A}^T \boldsymbol{x}$.

The following result shows, how the second derivatives of a function of n variables are transformed if a coordinate transformation by a rotation is made.

Lemma 7 *Let be given a twice continuously differentiable function $f : \mathbb{R}^n \to \mathbb{R}$ and an orthogonal $n \times n$-matrix $\boldsymbol{A} = (\boldsymbol{a}_1, \ldots, \boldsymbol{a}_n)$. A coordinate transformation is given by $\boldsymbol{y} = \boldsymbol{A}^T \boldsymbol{x}$. For the second derivatives of the function in the new coordinates $\tilde{f}(\boldsymbol{y}) = f(\boldsymbol{A}\boldsymbol{y})$ we have the following transformation formula*

$$\boldsymbol{H}_{\tilde{f}}(\boldsymbol{y}) = \boldsymbol{A}^T \boldsymbol{H}_f(\boldsymbol{A}\boldsymbol{y})\boldsymbol{A}. \tag{2.10}$$

PROOF: The gradient of $\tilde{f}(\boldsymbol{y})$ is

$$\nabla \tilde{f}(\boldsymbol{y}) = \boldsymbol{A}^T \nabla f(\boldsymbol{A}\boldsymbol{y}). \tag{2.11}$$

By differentiating further we get the second derivatives

$$\boldsymbol{H}_{\tilde{f}}(\boldsymbol{y}) = \boldsymbol{A}^T \boldsymbol{H}_f(\boldsymbol{A}\boldsymbol{y})\boldsymbol{A}. \tag{2.12}$$

□

In the terminology of tensors we have that the first derivatives of a function are a covariant first-order tensor and the second derivatives a covariant second-order tensor (see for example [83], p. 67/8).

Theorem 8 (Lebesgue Convergence Theorem) *Let $f(\boldsymbol{x}), f_1(\boldsymbol{x}), f_2(\boldsymbol{x}), \ldots$ be measurable functions on a measurable subset $D \subset \mathbb{R}^n$ such that*

1. $\lim_{m\to\infty} f_m(\boldsymbol{x}) = f(\boldsymbol{x})$ *almost everywhere.*
2. *There exists an integrable function $g(\boldsymbol{x})$ such that for all $\boldsymbol{x} \in D$ and for $m = 1, 2, \ldots$ always $|f_m(\boldsymbol{x})| \leq g(\boldsymbol{x})$.*

Then

$$\int_D f(\boldsymbol{x})\, d\boldsymbol{x} = \lim_{m\to\infty} \int_D f_m(\boldsymbol{x})\, d\boldsymbol{x}. \tag{2.13}$$

PROOF: See [62], p. 232. □

Theorem 9 (Implicit Function Theorem) *Let $U \subset \mathbb{R}^n$ and $V \subset \mathbb{R}^k$ be open sets and $(\boldsymbol{x}_0, \boldsymbol{y}_0) \in U \times V$. Further $f : U \times V \to \mathbb{R}^k$ is a C^k-vector function with $f(\boldsymbol{x}_0, \boldsymbol{y}_0) = \mathbf{o}$. If then*

$$\det\left(\left(\frac{\partial f_i(\boldsymbol{x}_0, \boldsymbol{y}_0)}{\partial y_j}\right)_{i,j=1,\ldots,k}\right) \neq 0, \tag{2.14}$$

then exists an open neighborhood $U_1 \subset U$ of $\boldsymbol{x}_0$ and a C^k-vector function $h: U_1 \rightarrow \mathbb{R}^k$ such that for $\boldsymbol{x} \in U_1$ always

$$f(\boldsymbol{x}, h(\boldsymbol{x})) = \mathbf{o} \text{ and } h(\boldsymbol{x}_0) = \boldsymbol{y}_0. \tag{2.15}$$

PROOF: See for example [102], p. 74/5. □

Theorem 10 (Morse Lemma) *Let $\boldsymbol{x}_0$ be a non-degenerate stationary point of the real-valued C^∞-function $f(\boldsymbol{x})$. Then there is a neighborhood U of $\boldsymbol{x}_0$ and a neighborhood V of $\mathbf{o}$ such that there is a C^∞-diffeomorphism $\boldsymbol{T}: U \rightarrow V$ with*

$$f(\boldsymbol{T}^{-1}(\boldsymbol{y})) = f(\boldsymbol{x}_0) + \frac{1}{2}\boldsymbol{y}^T \boldsymbol{H}_f(\boldsymbol{x}_0)\boldsymbol{y}, \tag{2.16}$$

and such that for the transformation determinant $J(\boldsymbol{v})$ of the diffeomorphism $J(\mathbf{o}) = 1$.

PROOF: See for example [140], p. 479. □

If the Hessian is positive or negative definite, by a further linear transformation it can be achieved that the function or its negative can be written as the sum of the squares of the coordinates. We will need only the following simpler form of this result.

Corollary 11 *Let $f: F \rightarrow \mathbb{R}^n$ be a twice continously differentiable function, which has a the point $\boldsymbol{x}_0$ in the interior of F a non-degenerate maximum. Then there there is a neighborhood U of $\boldsymbol{x}_0$ and a neighborhood V of $\mathbf{o}$ such that there is a C^2-diffeomorphism $\boldsymbol{T}: U \rightarrow V$ with*

$$f(\boldsymbol{T}^{-1}(\boldsymbol{y})) = f(\boldsymbol{x}_0) - \frac{1}{2}|\boldsymbol{y}|^2. \tag{2.17}$$

Here the Jacobian $J(\boldsymbol{x})$ of the transformation has the value $|\det(H_f(\boldsymbol{x}_0))|^{1/2}$ at the point $\boldsymbol{x}_0$.

PROOF: We assume that $\boldsymbol{x}_0 = \mathbf{o}$ and $f(\mathbf{o}) = 0$. Making a suitable rotation, we can always achieve that the unit vectors $\boldsymbol{e}_i$ are the eigenvectors of $\boldsymbol{H}_f(\mathbf{o})$ with negative eigenvalues $\lambda_1, \ldots, \lambda_n$. Let K be a sphere with $\mathbf{o}$ in its center so small that $K \subset F$ and the Hessian $\boldsymbol{H}_f(\boldsymbol{x})$ is negative definite throughout K. We define a vector function $\boldsymbol{T}: V \rightarrow \mathbb{R}^n$ by

$$\boldsymbol{T}(\boldsymbol{x}) = |2f(\boldsymbol{x})|^{1/2} \frac{\boldsymbol{x}}{|\boldsymbol{x}|} \tag{2.18}$$

with $\boldsymbol{T}(\boldsymbol{x}) = (t_1(\boldsymbol{x}), \ldots, t_n(\boldsymbol{x}))$. For $\boldsymbol{x} = \mathbf{o}$ we set $\boldsymbol{T}(\mathbf{o}) = \mathbf{o}$. Then we have $-|\boldsymbol{T}(\boldsymbol{x})|^2/2 = f(\boldsymbol{x})$.

From the definitions follows that this vector function is a C^2-function. We show that T is an one-to-one mapping in a subset of K with the origin in its interior. Since we have for the function $f(\boldsymbol{x})$ near $\mathbf{o}$ the expansion

$$f(\boldsymbol{x}) = \frac{1}{2}\sum_{i=1}^{n} \lambda_i x_i^2 + o(|\boldsymbol{x}|^2), \tag{2.19}$$

we get for the vector function the expansion

$$\boldsymbol{T}(\boldsymbol{x}) = -\left(\frac{\sum_{i=1}^{n} \lambda_i x_i^2}{\sum_{i=1}^{n} x_i^2}\right)^{1/2} \boldsymbol{x} + o(|\boldsymbol{x}|^2). \tag{2.20}$$

From this we get for the first derivatives

$$\frac{\partial t_j}{\partial x_i} = -\lambda_i^{1/2}\delta_{ij} + o(|\boldsymbol{x}|),\ \boldsymbol{x} \to \mathbf{o} \tag{2.21}$$

and therefore we find for the Jacobian

$$J(\boldsymbol{x}) = -\prod_{i=1}^{n} \lambda_i^{1/2} + o(|\boldsymbol{x}|),\ \boldsymbol{x} \to \mathbf{o}. \tag{2.22}$$

So we have $J(\mathbf{o}) = -\prod_{i=1}^{n} \lambda_i^{1/2} = |\det(\boldsymbol{H}_f(\mathbf{o}))|^{1/2}$.

Therefore the Jacobian is not zero at $\mathbf{o}$ and there exists a neighborhood $U \subset K$ of $\mathbf{o}$ that the vector function is an one-to-one mapping and defines a C^2-diffeomorphism.

Near $\mathbf{o}$ we have that

$$t_j \sim -\lambda_j^{1/2} x_j + o(|\boldsymbol{x}|) \tag{2.23}$$

and

$$x_i \sim -\lambda_i^{-1/2} t_i + o(|\boldsymbol{T}(\boldsymbol{x})|). \tag{2.24}$$

□

2.3 Manifolds

Here some basic results about manifolds are collected. More detailed descriptions can be found in [62], [102] and [130].

Definition 3 *Let $1 \leq r < n$ and $q \geq 1$. A non-empty set $M \subset \mathbb{R}^n$ is called an r-dimensional C^q-manifold if for each $\boldsymbol{x}_0 \in M$ there exists a neighborhood U of $\boldsymbol{x}_0$ and a C^q-vector function $\Psi : U \to \mathbb{R}^{n-r}$ with $\Psi = (\Psi^1, \dots, \Psi^{n-r})$ such that $D\Psi(\boldsymbol{x}) = (\nabla\Psi^1(\boldsymbol{x}), \dots, \nabla\Psi^{n-r}(\boldsymbol{x}))$ has rank $n - r$ for all $\boldsymbol{x} \in U$ and*

$$M \cap U = \{\boldsymbol{x} \in U; \Psi(\boldsymbol{x}) = \mathbf{o}\}.$$

An $(n-1)$-dimensional manifold in the n-dimensional space is usually called an hypersurface.

Definition 4 *Let g be a function from an open set $D \subset \mathbb{R}^r$ into an r-dimensional manifold $M \subset \mathbb{R}^n$ with $r \leq n$. Then g is called a regular transformation if*

1. *g is a C^1-vector function,*
2. *g is injective,*
3. *The matrix $Dg(\boldsymbol{t}) = (\nabla g_1(\boldsymbol{t}), \dots, \nabla g_n(\boldsymbol{t}))$ has rank r for all $\boldsymbol{t} \in D$.*

Definition 5 *Let S be a non-empty, relativly open subset of an r-dimensional C^q-manifold. A bijective transformation $\boldsymbol{F}$ of S in $\mathbb{R}^r$ is called a local coordinate system for S if $\boldsymbol{F} = \boldsymbol{T}^{-1}$ and $\boldsymbol{T}$ is a regular transformation from an open set $\boldsymbol{\Delta} \subset \mathbb{R}^r$ on S.*

This means that each point $\boldsymbol{x} \in S$ is uniquely determined by the values $F_1(\boldsymbol{x}), \ldots, F_r(\boldsymbol{x})$. For example on a sphere in $\mathbb{R}^3$ a point is determined by its latitude and altitude.

On an $(n-1)$-dimensional manifold $G = \{\boldsymbol{x}; g(\boldsymbol{x}) = 0\}$ defined by a function $g : \mathbb{R}^n \to \mathbb{R}$ with $\nabla g(\boldsymbol{x}) \neq \boldsymbol{o}$ for all $\boldsymbol{x} \in G$, a normal vector field can be defined by

$$\boldsymbol{n}(\boldsymbol{x}) = |\nabla g(\boldsymbol{x})|^{-1} \nabla g(\boldsymbol{x}). \tag{2.25}$$

This normal field defines an orientation on G. The tangential space of a manifold M at a point $\boldsymbol{x} \in M$ will be denoted by $T_M(\boldsymbol{x})$.

The Weingarten mapping defined in the following describes how a surface is curved. A morer precise description of the Weingarten mapping can be found in [130], p. 55/6.

Definition 6 *The Weingarten mapping of an $(n-1)$-dimensional C^1-manifold G in $\mathbb{R}^n$ oriented by the normal vector field $\boldsymbol{n}(\boldsymbol{x})$ at a point $\boldsymbol{x} \in G$ is defined by*

$$\begin{aligned} L_{n-1} : T_G(\boldsymbol{x}) \to T_G(\boldsymbol{x}), \boldsymbol{v} \mapsto L_{n-1}(\boldsymbol{v}) \qquad (2.26) \\ = \left(-\sum_{i=1}^{n} v_i \cdot \frac{\partial n_1(\boldsymbol{x})}{\partial x_i}, \ldots, -\sum_{i=1}^{n} v_i \cdot \frac{\partial n_n(\boldsymbol{x})}{\partial x_i} \right). \end{aligned}$$

The Weingarten mapping gives a measure for the change of the direction of the normal vector on the surface G if one moves through the point $\boldsymbol{x}$ with velocity vector $\boldsymbol{v}$ at $\boldsymbol{x}$. Since the Weingarten mapping is a linear self-adjoint mapping it has $n-1$ eigenvectors $\boldsymbol{v}_1, \ldots, \boldsymbol{v}_{n-1}$ with eigenvalues $\kappa_1, \ldots, \kappa_{n-1}$ (see for example [130], p. 86). The $\boldsymbol{v}_i$'s are called main (or principal) curvature directions and the κ_i's the main (principal) curvatures.

The curvature $\kappa(\boldsymbol{v})$ of a curve through $\boldsymbol{x}$ in G in direction of $\boldsymbol{v}$ is given by (see [130], p. 92)

$$\kappa(\boldsymbol{v}) = -\frac{1}{|\nabla g(\boldsymbol{x})|} \sum_{i,j=1}^{n} \frac{\partial^2 g(\boldsymbol{x})}{\partial x_i \partial x_j} v_i v_j. \tag{2.27}$$

This definition of the curvature is such that a sphere around the origin has positive curvature.

Since the Weingarten map is linear, it can be described by a matrix. The determinant and the trace of the Weingarten map are important in differential geometry. The determinant $K(p) = \det(L_p)$ is called the Gauss-Kronecker curvature of the surface at $\boldsymbol{x}$. It is equal to the product of the main curvatures, i.e. $K(p) = \prod_{i=1}^{n-1} \kappa_i$. n^{-1} times the trace of L_p is called the mean curvature $H(p)$ of the surface at $\boldsymbol{x}$. Therefore $H(p) = n^{-1} \sum_{i=1}^{n-1} \kappa_i$ is the average of the main curvatures.

The volume $\alpha_n(\rho)$ of an n-dimensional sphere $\{\boldsymbol{x}; |\boldsymbol{x}| \leq \rho\}$ is given by (see [62], p. 218)

$$\alpha_n(\rho) = \frac{2 \cdot \pi^{n/2}}{n\Gamma(n/2)} \cdot \rho^n \tag{2.28}$$

and the $(n-1)$-dimensional surface $\beta_n(\rho)$ of this sphere is found by differentiating this expression with respect to ρ giving

$$\beta_n(\rho) = \frac{2 \cdot \pi^{n/2}}{\Gamma(n/2)} \cdot \rho^{n-1}. \tag{2.29}$$

Similar as in the univariate case there are relations between the derivatives of a vector function in a domain and the integral of the function over the boundary.

Theorem 12 (Divergence Theorem) *Let G be a compact $(n-1)$-dimensional submanifold in the n-dimensional space, defined by a C^1-function $g : \mathbb{R}^n \to \mathbb{R}$ by $G = \{\boldsymbol{x}; g(\boldsymbol{x}) = 0\}$ such that the gradient $\nabla g(\boldsymbol{y})$ does not vanish on G and that $F = \{\boldsymbol{x}; g(\boldsymbol{x}) \leq 0\}$ is a compact domain. Then, if the manifold is oriented by the normal field $\boldsymbol{n}(\boldsymbol{y}) = |\nabla g(\boldsymbol{y})|^{-1}\nabla g(\boldsymbol{y})$, for a continuously differentiable vector field $\boldsymbol{u}(\boldsymbol{x})$ the following equation holds*

$$\int_F \operatorname{div}(\boldsymbol{u}(\boldsymbol{x}))\, d\boldsymbol{x} = \int_G \langle \boldsymbol{n}(\boldsymbol{y}), \boldsymbol{u}(\boldsymbol{y})\rangle\, ds(\boldsymbol{y}). \tag{2.30}$$

Here $ds(\boldsymbol{y})$ denotes surface integration over G.

PROOF: See for example [102], p. 319. □

2.4 Extrema under Constraints

Here some results about extrema under constraints are given (For more details see [70]).

Definition 7 *Given is a set $U \subset \mathbb{R}^n$ and a subset $M \subset U$. A function $f : U \to \mathbb{R}$ has at a point $\boldsymbol{x} \in M$ a local maximum (minimum) under the constraint M if there exists a $\delta > 0$ such that for $\boldsymbol{z} \in M$ with $|\boldsymbol{z} - \boldsymbol{x}| \leq \delta$ always*

$$f(\boldsymbol{z}) \leq f(\boldsymbol{x}) \text{ (resp. } f(\boldsymbol{z}) \geq f(\boldsymbol{x})).$$

Theorem 13 (Lagrange Multiplier Rule) *Let $U \subset \mathbb{R}^n$ be an open set. On this set let be defined the continuously differentiable functions $f : U \to \mathbb{R}$ and $g_1, \ldots, g_k : U \to \mathbb{R}$ such that an $(n-k)$-dimensional C^1-manifold $M \subset U$ is defined by*

$$M = \{\boldsymbol{x}; g_1(\boldsymbol{x}) = \ldots = g_k(\boldsymbol{x}) = 0\}.$$

If the function f has a local extremum under the constraint M at a point $\boldsymbol{x}^* \in M$ and the gradients $\nabla g_1(\boldsymbol{x}^*), \ldots, \nabla g_k(\boldsymbol{x}^*)$ are linearly independent, there exist numbers $\lambda_1, \ldots, \lambda_k$ such that

$$\nabla f(\boldsymbol{x}^*) = \sum_{i=1}^{k} \lambda_i \nabla g_i(\boldsymbol{x}^*).$$

These numbers are called Lagrange multipliers.

PROOF: See [62], p. 161. □

If the coordinate system is changed to local coordinates on a surface, the form of the second derivatives with respect to these coordinates at a local extremum on a manifold is derived in the following lemma.

Lemma 14 *Let be given a twice continuously differentiable function $f : \mathbb{R}^n \to \mathbb{R}$ and k twice continuously differentiable functions $g_i : \mathbb{R}^n \to \mathbb{R}$ with $k \leq n$. The functions g_i define an $(n-k)$-dimensional manifold $G = \cap_{i=1}^{k}\{\boldsymbol{x}; g_i(\boldsymbol{x}) = 0\}$. Assume that further:*

1. *The function f has at the point $\boldsymbol{x}^* \in G$ a local maximum (minimum) with respect to G and the gradients $\nabla g_1(\boldsymbol{x}^*), \ldots, \nabla g_k(\boldsymbol{x}^*)$ are linearly independent.*

2. *The gradient $\nabla f(\boldsymbol{x}^*)$ can be written as*

$$\nabla f(\boldsymbol{x}^*) = \sum_{l=1}^{k} \gamma_l \nabla g_l(\boldsymbol{x}^*). \tag{2.31}$$

   (This follows from 1. using the Lagrange multiplier theorem 13, p. 16.)

3. *In a neighborhood of $\boldsymbol{x}^*$ in G there is a local coordinate system given by the coordinates $u_1, \ldots, u_{n-k}$ defined by the inverse of the function $T : U \to G, (u_1, \ldots, u_{n-k}) \mapsto (x_1(\boldsymbol{u}), \ldots, x_n(\boldsymbol{u}))$.*

Then the second partial derivatives of the function $f(T(u_1, \ldots, u_{n-k}))$ with respect to the local coordinates $u_1, \ldots, u_{n-k}$ have the form

$$\left(\frac{\partial^2 f(T(u_1, \ldots, u_{n-k}))}{\partial u_i \partial u_j}\right)_{i,j=1,\ldots,n-k} = \boldsymbol{D} \cdot \boldsymbol{H}(\boldsymbol{x}^*) \cdot \boldsymbol{D}^T, \tag{2.32}$$

where

1.
$$\boldsymbol{D} = (\nabla x_1(\boldsymbol{u}), \ldots, \nabla x_n(\boldsymbol{u})) = \left(\frac{\partial x_j}{\partial u_i}\right)_{i,j=1,\ldots,n} \tag{2.33}$$

   is the Jacobian of the regular transformation to local coordinates.

2.
$$\boldsymbol{H}(\boldsymbol{x}^*) = \left(f^{ij}(\boldsymbol{x}^*) - \sum_{l=1}^{k} \gamma_l g_l^{ij}(\boldsymbol{x}^*)\right)_{i,j=1,\ldots,n}. \tag{2.34}$$

PROOF: First we calculate the first partial derivatives of $f(T(u_1, \ldots, u_{n-k}))$. They are

$$\frac{\partial f}{\partial u_i} = \sum_{r=1}^{n} f^r(\boldsymbol{x}^*) \frac{\partial x_r}{\partial u_i} \text{ for } i = 1, \ldots, n-k. \tag{2.35}$$

The second derivatives with respect to the coordinates $u_1, \ldots, u_{n-k}$ are then

$$\frac{\partial^2 f}{\partial u_i \partial u_j} = \sum_{m,r=1}^{n} f^{mr}(\boldsymbol{x}^*) \frac{\partial x_r}{\partial u_i} \frac{\partial x_m}{\partial u_j} + \sum_{r=1}^{n} f^r(\boldsymbol{x}^*) \frac{\partial^2 x_r}{\partial u_i \partial u_j}. \tag{2.36}$$

Since the $u_1, \ldots, u_{n-k}$ are a local coordinate system for the manifold G, $g_l(T(u_1, \ldots, u_{n-k})) = 0$ for $l = 1, \ldots, k$. By differentiating this identity we get

$$\frac{\partial g_l(T(u_1, \ldots, u_{n-k}))}{\partial u_i} = \sum_{r=1}^{n} g_l^r(\boldsymbol{x}^*) \frac{\partial x_r}{\partial u_i} = 0. \tag{2.37}$$

Differentiating further gives

$$\frac{\partial^2 g_l(T(u_1, \ldots, u_{n-k}))}{\partial u_i \partial u_j} = \sum_{m,r=1}^{n} g_l^{mr}(\boldsymbol{x}^*) \frac{\partial x_r}{\partial u_i} \frac{\partial x_m}{\partial u_j} + \sum_{r=1}^{n} g_l^r(\boldsymbol{x}^*) \frac{\partial^2 x_r}{\partial u_i \partial u_j} = 0. \tag{2.38}$$

From the last equation we obtain then

$$\sum_{m,r=1}^{n} g_l^{mr}(\boldsymbol{x}^*) \frac{\partial x_r}{\partial u_i} \frac{\partial x_m}{\partial u_j} = -\sum_{r=1}^{n} g_l^r(\boldsymbol{x}^*) \frac{\partial^2 x_r}{\partial u_i \partial u_j}. \tag{2.39}$$

Now using equation (2.31) gives finally

$$\begin{aligned} \sum_{r=1}^{n} f^r(\boldsymbol{x}^*) \frac{\partial^2 x_r}{\partial u_i \partial u_j} &= \sum_{r=1}^{n} \sum_{l=1}^{k} \gamma_l g_l^r(\boldsymbol{x}^*) \frac{\partial^2 x_r}{\partial u_i \partial u_j} \\ &= -\sum_{l=1}^{k} \gamma_l \sum_{m,r=1}^{n} g_l^{mr}(\boldsymbol{x}^*) \frac{\partial x_r}{\partial u_i} \frac{\partial x_m}{\partial u_j}. \end{aligned} \tag{2.40}$$

Inserting this result in equation (2.36) gives the final result. □

Definition 8 *Let a set F be defined by*

$$g_j(\boldsymbol{x}) \leq 0 \text{ for } j = 1, \ldots m.$$

A point $\boldsymbol{x} \in F$ with $g_1(\boldsymbol{x}) = \ldots = g_k(\boldsymbol{x}) = 0$ for a number k with $1 \leq k \leq m$ and $g_j(\boldsymbol{x}) < 0$ for $j = k+1, \ldots, m$ is called a regular point of F if the gradients $\nabla g_1(\boldsymbol{x}), \ldots, \nabla g_k(\boldsymbol{x})$ are linearly independent.

For this general case of inequality constraints we have the following theorem.

Theorem 15 *Let be given a continuously differentiable function $f : \mathbb{R}^n \to \mathbb{R}$ and m continuously differentiable functions $g_i : \mathbb{R}^n \to \mathbb{R}$. If the function f has at the point $\boldsymbol{x}^*$ a local maximum (minimum) under the constraints*

$$g_j(\boldsymbol{x}) \leq 0 \text{ for } j = 1, \ldots m \tag{2.41}$$

and $\boldsymbol{x}^*$ is a regular point of $F = \cap_{j=1}^m \{g_j(\boldsymbol{x}) \leq 0\}$, there exist numbers $\lambda_1, \ldots, \lambda_m$ with $\lambda_j \geq 0$ (resp. $\lambda_j \leq 0$) for $j = 1, \ldots, k$ such that

$$\nabla f(\boldsymbol{x}^*) = \sum_{j=1}^{m} \lambda_j \nabla g_j(\boldsymbol{x}^*). \tag{2.42}$$

Here $\lambda_j = 0$ if $g_j(\boldsymbol{x}^*) < 0$ for $j = 1, \ldots, m$.

PROOF: See [70], p. 186. □

A sufficient condition for local extrema under constraints can be found by studying the second derivatives of the objective function and the constraint functions at a stationary point of the Lagrangian.

Definition 9 *Let $\boldsymbol{A}$ be a symmetric $n \times n$-matrix and $\boldsymbol{B}$ an $m \times n$-matrix with $m < n$. The matrix $\boldsymbol{A}$ is called positive (negative) definite under the constraint $\boldsymbol{Bx} = \mathbf{o}$ if for an $\boldsymbol{x} \in \mathbb{R}^n$ with $\boldsymbol{x} \neq \mathbf{o}$ and $\boldsymbol{Bx} = \mathbf{o}$ always*

$$\boldsymbol{x}^T \boldsymbol{A} \boldsymbol{x} > 0 \ (\text{resp. } \boldsymbol{x}^T \boldsymbol{A} \boldsymbol{x} < 0).$$

Theorem 16 *Let $U \subset \mathbb{R}^n$ be an open set. On this set are defined continuously differentiable functions $f : U \to \mathbb{R}$ and $g_1, \ldots, g_k : U \to \mathbb{R}$ such that an $(n-k)$-dimensional C^1-manifold $M \subset U$ is defined by*

$$M = \{\boldsymbol{x}; g_1(\boldsymbol{x}) = \ldots = g_k(\boldsymbol{x}) = 0\}. \tag{2.43}$$

For a local maximum (minimum) under the constraint M at a point $\boldsymbol{x}^*$ it is sufficient that the following conditions are fulfilled:

1. *The gradients $\nabla g_1(\boldsymbol{x}^*), \ldots, \nabla g_k(\boldsymbol{x}^*)$ are linearly independent and there are numbers $\lambda_1, \ldots, \lambda_k$ with*

$$\nabla f(\boldsymbol{x}^*) = \sum_{i=1}^{k} \lambda_i \nabla g_i(\boldsymbol{x}^*). \tag{2.44}$$

2. *The Hessian of the function $F(\boldsymbol{x}) = f(\boldsymbol{x}) + \sum_{i=1}^k \lambda_i g_i(\boldsymbol{x})$ at $\boldsymbol{x}^*$ is negative (resp. positive) definite under the constraint $\boldsymbol{Bx} = \mathbf{o}$ where $\boldsymbol{B} = (\nabla g_1(\boldsymbol{x}^*), \ldots, \nabla g_k(\boldsymbol{x}^*))^T$.*

PROOF: See [5], p. 213-214. □

The usual criterion for the definiteness of a matrix under linear constraints is the following. Here for a $k \times l$-matrix $\boldsymbol{C}$ the symbol $\boldsymbol{C}^{mi}$ denotes the $m \times i$-matrix consisting of the first m rows and i columns of the matrix $\boldsymbol{C}$.

Theorem 17 *Let $\boldsymbol{A}$ be a symmetric $n \times n$-matrix and $\boldsymbol{B}$ an $m \times n$-matrix with $m < n$ and $\mathrm{rk}(\boldsymbol{B}) = m$. Then:*

1. *$\boldsymbol{A}$ is positive definite under the constraint $\boldsymbol{Bx} = \mathbf{o}$ iff*

$$(-1)^m \det \begin{pmatrix} \boldsymbol{A}^{ii} & (\boldsymbol{B}^{mi})^T \\ \boldsymbol{B}^{mi} & 0 \end{pmatrix} > 0 \text{ for } i = m+1, \ldots, n.$$

2. *$\boldsymbol{A}$ is negative definite under the constraint $\boldsymbol{Bx} = \mathbf{o}$ iff*

$$(-1)^i \det \begin{pmatrix} \boldsymbol{A}^{ii} & (\boldsymbol{B}^{mi})^T \\ \boldsymbol{B}^{mi} & 0 \end{pmatrix} > 0 \text{ for } i = m+1, \ldots, n.$$

PROOF: A proof can be found in [93], p. 136. □

In the next theorem a new simpler condition is given. The idea here is to project the matrix onto the subspace, which is orthogonal to the subspace defined by $\boldsymbol{Bx} = \mathbf{o}$.

Theorem 18 *Let $\boldsymbol{A}$ be a symmetric $n \times n$-matrix and $\boldsymbol{B}$ an $m \times n$-matrix with $m < n$ and $\mathrm{rk}(\boldsymbol{B}) = m$. $U \subset \mathbb{R}^n$ is defined by $U = \{\boldsymbol{x}; \boldsymbol{Bx} = \mathbf{o}\}$. Then the following statements are equivalent:*
a) The matrix $\boldsymbol{A}$ is positive (negative) definite under the constraint $\boldsymbol{Bx} = \mathbf{o}$.
b) The $n \times n$-matrix $\boldsymbol{A}_+ = \boldsymbol{P}_U^T \boldsymbol{A} \boldsymbol{P}_U + \boldsymbol{P}_{U^\perp}^T \boldsymbol{P}_{U^\perp}$ ($\boldsymbol{A}_- = \boldsymbol{P}_U^T \boldsymbol{A} \boldsymbol{P}_U - \boldsymbol{P}_{U^\perp}^T \boldsymbol{P}_{U^\perp}$) is positive (negative) definite.

PROOF: We prove only the case "positive definite".
b) $\Rightarrow$ *a)*: If the matrix $\boldsymbol{A}_+$ is positive definite, we have for all $\boldsymbol{x} \neq \mathbf{o}$ with $\boldsymbol{Bx} = \mathbf{o}$ that

$$\begin{aligned} \boldsymbol{x}^T(\boldsymbol{P}_U^T \boldsymbol{A} \boldsymbol{P}_U + \boldsymbol{P}_{U^\perp}^T \boldsymbol{P}_{U^\perp})\boldsymbol{x} &= \boldsymbol{x}^T \boldsymbol{P}_U^T \boldsymbol{A} \boldsymbol{P}_U \boldsymbol{x} + \boldsymbol{x}^T \boldsymbol{P}_{U^\perp}^T \boldsymbol{P}_{U^\perp} \boldsymbol{x} \\ = \boldsymbol{x}^T \boldsymbol{A} \boldsymbol{x} + 0 &= \boldsymbol{x}^T \boldsymbol{A} \boldsymbol{x} > 0. \end{aligned} \tag{2.45}$$

This gives statement a).
a) $\Rightarrow$ *b)*: Let be $\boldsymbol{z} \in \mathbb{R}^n$, then it can be written in the form $\boldsymbol{z} = \boldsymbol{x} + \boldsymbol{y}$ with $\boldsymbol{x} \in U$ and $\boldsymbol{y} \in U^\perp$. Then we have that

$$\boldsymbol{z}^T \boldsymbol{A}_+ \boldsymbol{z} = (\boldsymbol{x} + \boldsymbol{y})^T (\boldsymbol{P}_U^T \boldsymbol{A} \boldsymbol{P}_U + \boldsymbol{P}_{U^\perp}^T \boldsymbol{P}_{U^\perp})(\boldsymbol{x} + \boldsymbol{y}), \tag{2.46}$$

and since $\boldsymbol{P}_{U^\perp} \boldsymbol{x} = \mathbf{o}$ and $\boldsymbol{P}_U \boldsymbol{y} = \mathbf{o}$, we get

$$= \boldsymbol{x}^T \boldsymbol{P}_U^T \boldsymbol{A} \boldsymbol{P}_U \boldsymbol{x} + \boldsymbol{y}^T \boldsymbol{P}_{U^\perp}^T \boldsymbol{P}_{U^\perp} \boldsymbol{y} = \boldsymbol{x}^T \boldsymbol{A} \boldsymbol{x} + |\boldsymbol{y}|^2. \tag{2.47}$$

Since $\boldsymbol{x} \in U$ from this follows that $\boldsymbol{x} \boldsymbol{A} \boldsymbol{x} > 0$ if $\boldsymbol{x} \neq \mathbf{o}$. Since $|\boldsymbol{y}|^2 > 0$ iff $\boldsymbol{y} \neq \mathbf{o}$, we get therefore that

$$\boldsymbol{z}^T \boldsymbol{A}_+ \boldsymbol{z} > 0 \tag{2.48}$$

iff $\boldsymbol{z} \neq \mathbf{o}$. But this means that the matrix $\boldsymbol{A}_+$ is positive definite. □

Using this criterion it is only necessary to check, if *one* $n \times n$-matrix is definite. The easiest way to do this is to make a Cholesky decomposition of the matrix (or of its negative), see [129], p. 145-8 and [139], p. 13. If the decomposition algorithm breaks down, the matrix is not positive definite.

Further this result can be used for calculating the determinant of matrices, which are obtained by projection onto subspaces.

Corollary 19 *Given is an $n \times n$-matrix $\boldsymbol{B}$ and a k-dimensional subspace $U \subset \mathbb{R}^n$ defined by $U = \text{span}[\boldsymbol{a}_1, \ldots, \boldsymbol{a}_k]$, where the $\boldsymbol{a}_l$'s are orthonormal vectors. Then the determinant of the $k \times k$ matrix $\boldsymbol{A}^T \boldsymbol{B} \boldsymbol{A}$ is equal to the determinant of the $n \times n$ matrix $\boldsymbol{P}_U^T \boldsymbol{B} \boldsymbol{P}_U + (\boldsymbol{I}_n - \boldsymbol{P}_U)^T (\boldsymbol{I}_n - \boldsymbol{P}_U)$, i.e.*

$$\det(\boldsymbol{A}^T \boldsymbol{B} \boldsymbol{A}) = \det(\boldsymbol{P}_U^T \boldsymbol{B} \boldsymbol{P}_U + (\boldsymbol{I}_n - \boldsymbol{P}_U)^T (\boldsymbol{I}_n - \boldsymbol{P}_U)). \tag{2.49}$$

PROOF: We assume that the vectors $\boldsymbol{a}_i$ are the first k unit vectors, i.e. $\boldsymbol{a}_i = \boldsymbol{e}_i$ for $i = 1, \ldots, k$. Then we get

$$\boldsymbol{A}^T \boldsymbol{B} \boldsymbol{A} = (b_{ij})_{i,j=1,\ldots,k} = \tilde{\boldsymbol{B}}, \tag{2.50}$$

and that

$$\boldsymbol{P}_U = \begin{pmatrix} \boldsymbol{I}_k & \mathbf{o}_{k,n-k} \\ \mathbf{o}_{n-k,k} & \mathbf{o}_{n-k,n-k} \end{pmatrix}. \tag{2.51}$$

Then we have

$$\boldsymbol{P}_U^T \boldsymbol{B} \boldsymbol{P}_U + (\boldsymbol{I}_n - \boldsymbol{P}_U)^T (\boldsymbol{I}_n - \boldsymbol{P}_U) = \begin{pmatrix} \tilde{\boldsymbol{B}} & \mathbf{o}_{k,n-k} \\ \mathbf{o}_{n-k,k} & \boldsymbol{I}_{n-k} \end{pmatrix}. \tag{2.52}$$

But the determinant of this matrix is equal to the determinant of $\tilde{\boldsymbol{B}} = (b_{ij})_{i,j=1,\ldots,k}$. The general case is proved by making a rotation. □

2.5 Parameter Dependent Integrals

The derivative of an integral where integrand and limits depending on a parameter τ is given by the Leibniz theorem. Then the Leibniz theorem gives the derivative of such an integral with respect to this parameter (see [1], p. 11, 3.3.7).

Theorem 20 (Leibniz Theorem) *Let $I \subset \mathbb{R}$ be an open interval. Given are two twice continuously differentiable functions $a : I \to \mathbb{R}$ and $b : I \to \mathbb{R}$, and a continuous function $\mathbb{R} \times I \to \mathbb{R}$, which is continuously partially differentiable with respect to the second argument τ. Then the integral*

$$I(\tau) = \int_{a(\tau)}^{b(\tau)} f(x,\tau)\, dx \tag{2.53}$$

is differentiable with respect to τ and its derivative is given by

$$\begin{aligned} I'(\tau) &= \frac{d}{d\tau} \int_{a(\tau)}^{b(\tau)} f(x,\tau)\, dx \\ &= \int_{a(\tau)}^{b(\tau)} \frac{\partial f(x,\tau)}{\partial \tau}\, dx + \left[b'(\tau) f(b(\tau),\tau) - a'(\tau) f(a(\tau),\tau)\right]. \end{aligned} \tag{2.54}$$

The first summand describes the influence of τ on the function $f(x,\tau)$ and the term in the square brackets the influence on the boundary points.

PROOF: We prove only the case that $a(\tau)$ is identically equl to zero, i.e. $a(\tau) \equiv 0$. The general case follows easily. We have

$$\begin{aligned} I(\tau+h) - I(\tau) &= \int_0^{b(\tau)} \left[f(x,\tau+h) - f(x,\tau)\right]\, dx \\ &+ \int_{b(\tau)}^{b(\tau+h)} f(x,\tau+h)\, dx. \end{aligned} \tag{2.55}$$

Using the mean value theorem, this can be written as

$$I(\tau+h) - I(\tau) = \int_0^{b(\tau)} h f(x,\tau + h\vartheta(x))\, dx \tag{2.56}$$

$$+(b(\tau+h) - b(\tau)) f(x + \vartheta^*(b(\tau+h) - b(\tau)), \tau + h).$$

Here $0 \le \vartheta(x) \le 1$ and $0 \le \vartheta^* \le 1$. This gives then for the difference quotient

$$\frac{I(\tau+h) - I(\tau)}{h} = \int_0^{b(\tau)} f(x,\tau + h\vartheta(x))\, dx \tag{2.57}$$

$$+\frac{b(\tau+h) - b(\tau)}{h} f(x + \vartheta^*(b(\tau+h) - b(\tau)), \tau + h).$$

For the first integral we get using the Lebesgue theorem that

$$\int_0^{b(\tau)} f(x, \tau + h\vartheta(x))\, dx \to \int_0^{b(\tau)} f(x, \tau)\, dx,\ h \to 0, \tag{2.58}$$

and for the second term we have as $h \to 0$ that

$$\frac{b(\tau + h) - b(\tau)}{h} f(x + \vartheta^*(b(\tau + h) - b(\tau)), \tau + h) \to b'(\tau)f(\tau). \tag{2.59}$$

This gives the result in this case. □

This result is generalized to functions of several variables in the following theorem. Here the boundary of the integration domain is a surface in $\mathbb{R}^n$.

Theorem 21 *Let be given two continuously differentiable functions* $f : \mathbb{R}^n \times I \to \mathbb{R}, (\boldsymbol{x}, \tau) \mapsto f(\boldsymbol{x}, \tau)$ *and* $g : \mathbb{R}^n \times I \to \mathbb{R}, (\boldsymbol{x}, \tau) \mapsto g(\boldsymbol{x}, \tau)$ *with* I *an open interval. Assume further that:*

1. *There exist integrable functions* $h, h_1 : \mathbb{R}^n \to \mathbb{R}$ *such that for* $\tau \in I$ *always*

$$|f(\boldsymbol{x}, \tau)| < h(\boldsymbol{x}) \text{ and } |f_\tau(\boldsymbol{x}, \tau)| < h_1(\boldsymbol{x}). \tag{2.60}$$

2. *For all* $\tau \in I$ *the set* $G(\tau) = \{\boldsymbol{x}; g(\boldsymbol{x}, \tau) = 0\}$ *is a compact* C^1*-manifold in* $\mathbb{R}^n$ *and for all* $\boldsymbol{x} \in G(\tau)$ *always* $\nabla_{\boldsymbol{x}} g(\boldsymbol{x}, \tau) \neq \mathbf{o}$. *(An orientation for* $G(\tau)$ *is given by the normal field* $\boldsymbol{n}_\tau(\boldsymbol{x}) = |\nabla_{\boldsymbol{x}} g(\boldsymbol{x}, \tau)|^{-1} \nabla_{\boldsymbol{x}} g(\boldsymbol{x}, \tau)$ *).*

Then the integral

$$F(\tau) = \int_{g(\boldsymbol{x}, \tau) \leq 0} f(\boldsymbol{x}, \tau)\, d\boldsymbol{x} \tag{2.61}$$

exists for all $\tau \in I$ *and the derivative of this integral with respect to* τ *is given by*

$$F'(\tau) = \int_{g(\boldsymbol{x}, \tau) \leq 0} f_\tau(\boldsymbol{x}, \tau)\, d\boldsymbol{x} - \int_{G(\tau)} f(\boldsymbol{y}, \tau) \frac{g_\tau(\boldsymbol{y}, \tau)}{|\nabla_{\boldsymbol{y}} g(\boldsymbol{y}, \tau)|}\, ds_\tau(\boldsymbol{y}). \tag{2.62}$$

Here $ds_\tau(\boldsymbol{y})$ *denotes surface integration over* $G(\tau)$.

PROOF: In the following a short proof for this result is given. The existence of the integrals follows from the first condition. To derive the form of the derivative, we first write the difference $F(\tau + h) - F(\tau)$ as the sum of two terms

$$F(\tau + h) - F(\tau) = \int_{g(\boldsymbol{x}, \tau + h) \leq 0} f(\boldsymbol{x}, \tau + h)\, d\boldsymbol{x} - \int_{g(\boldsymbol{x}, \tau) \leq 0} f(\boldsymbol{x}, \tau)\, d\boldsymbol{x} \tag{2.63}$$

$$= \underbrace{\int\limits_{g(\boldsymbol{x},\tau+h)\le 0} (f(\boldsymbol{x},\tau+h)-f(\boldsymbol{x},\tau))\,d\boldsymbol{x}}_{=:D_1(h)}$$

$$+ \underbrace{\int\limits_{g(\boldsymbol{x},\tau+h)\le 0} f(\boldsymbol{x},\tau)\,d\boldsymbol{x} - \int\limits_{g(\boldsymbol{x},\tau)\le 0} f(\boldsymbol{x},\tau)\,d\boldsymbol{x}}_{=:D_2(h)}.$$

For small enough $\epsilon > 0$ in a neighborhood $G_\epsilon(\tau) = \{\boldsymbol{x}; \min_{\boldsymbol{y}\in G(\tau)} |\boldsymbol{y}-\boldsymbol{x}| < \epsilon\}$ of $G(\tau)$ a coordinate system can be introduced where each $\boldsymbol{y} \in G_\epsilon(\tau)$ which is given in the form $\boldsymbol{y} = \boldsymbol{x} + \delta \cdot \boldsymbol{n}_\tau(\boldsymbol{x})$ (here $\boldsymbol{n}_\tau(\boldsymbol{x})$ denotes the surface normal at $\boldsymbol{x}$) with $\boldsymbol{x} \in G(\tau)$ has the coordinates $(\boldsymbol{x}, \delta)$.

In this coordinate system the difference $D_2(h)$ can be written in the form

$$D_2(h) = \int\limits_{g(\boldsymbol{x},\tau)=0} \left[\int\limits_0^{l(\boldsymbol{x},h)} f(\boldsymbol{x}+\delta\cdot\boldsymbol{n}_\tau(\boldsymbol{x}),\tau)D(\boldsymbol{x},\delta)\,d\delta \right] ds_\tau(\boldsymbol{x}). \tag{2.64}$$

Here $D(\boldsymbol{x},\delta)$ is the transformation determinant for the change of the coordinates.

Due to the definition of the coordinates we have that $D(\boldsymbol{x},0) = 1$. The C^1-function $l(\boldsymbol{x},h)$ is defined implicitly by the equation $g(\boldsymbol{x} + l(\boldsymbol{x},h)\boldsymbol{n}_\tau(\boldsymbol{x}), \tau + h) = 0$. The existence of such a C^1-function can be proved (for sufficiently small h) using the implicit function theorem (see theorem 9, p. 12) if always $\nabla_{\boldsymbol{y}} g(\boldsymbol{y},\tau) \neq \mathrm{o}$ for $\boldsymbol{y} \in G(\tau)$, which is assumed in condition 2. Using the mean value theorem, we get

$$\begin{aligned} & \int\limits_0^{l(\boldsymbol{x},h)} f(\boldsymbol{x}+\delta\cdot\boldsymbol{n}_\tau(\boldsymbol{x}),\tau)D(\boldsymbol{x},\delta)\,d\delta \\ = \; & l(\boldsymbol{x},h) f(\boldsymbol{x}+\vartheta(\boldsymbol{x})\delta\cdot\boldsymbol{n}_\tau(\boldsymbol{x}),\tau)D(\boldsymbol{x},\vartheta(\boldsymbol{x})\delta). \end{aligned} \tag{2.65}$$

with $0 \le \vartheta(\boldsymbol{x}) \le 1$. Making a Taylor expansion of $l(\boldsymbol{x},0)$ with respect to h gives

$$l(\boldsymbol{x},h) = h l_h(\boldsymbol{x},0) + o(h). \tag{2.66}$$

Inserting this in equation (2.65) and then integrating gives in the limit as $h \to 0$ that

$$\lim_{h\to 0} \frac{1}{h} D_2(h) = \int\limits_{g(\boldsymbol{x},\tau)=0} l_h(\boldsymbol{x},0) f(\boldsymbol{x},\tau)\,ds_\tau(\boldsymbol{x}). \tag{2.67}$$

Making a Taylor expansion of the function $g(\boldsymbol{x},\tau)$ we get

$$\begin{aligned} & g(\boldsymbol{x}+l(\boldsymbol{x},h)\boldsymbol{n}_\tau(\boldsymbol{x}),\tau+h) - g(\boldsymbol{x},\tau) \\ = \; & l(\boldsymbol{x},h)\langle \boldsymbol{n}_\tau(\boldsymbol{x}), \nabla_{\boldsymbol{x}} g(\boldsymbol{x})\rangle + h g_\tau(\boldsymbol{x},\tau) + o(h) \end{aligned} \tag{2.68}$$

and using again the expansion in equation (2.66) of $l(\boldsymbol{x}, h)$ with respect to h yields further

$$= hl_h(\boldsymbol{x}, 0)\langle \boldsymbol{n}_\tau(\boldsymbol{x}), \nabla_{\boldsymbol{x}} g(\boldsymbol{x})\rangle + hg_\tau(\boldsymbol{x}, \tau) + o(h). \tag{2.69}$$

Since $g(\boldsymbol{x} + l(\boldsymbol{x}, h)\boldsymbol{n}_\tau(\boldsymbol{x}), \tau + h) = 0$ and $\langle \boldsymbol{n}_\tau(\boldsymbol{x}), \nabla_{\boldsymbol{x}} g(\boldsymbol{x}, \tau)\rangle = |\nabla_{\boldsymbol{x}} g(\boldsymbol{x}, \tau)|$, we get from the last equation that

$$h \cdot l_h(\boldsymbol{x}, 0)|\nabla_{\boldsymbol{x}} g(\boldsymbol{x}, \tau)| = -h \cdot g_\tau(\boldsymbol{x}, \tau) + o(h). \tag{2.70}$$

As $h \to 0$ this gives in the limit

$$l_h(\boldsymbol{x}, 0) = -g_\tau(\boldsymbol{x}, \tau)|\nabla_{\boldsymbol{x}} g(\boldsymbol{x}, \tau)|^{-1}. \tag{2.71}$$

This gives for $D_2(h)$ then

$$\lim_{h \to 0} \frac{1}{h} D_2(h) = - \int\limits_{g(\boldsymbol{x}, \tau)=0} g_\tau(\boldsymbol{x}, \tau)|\nabla_{\boldsymbol{x}} g(\boldsymbol{x}, \tau)|^{-1} f(\boldsymbol{x}, \tau)\, ds_\tau(\boldsymbol{x}). \tag{2.72}$$

For the first integral $D_1(h)$ we obtain, since g is a continuous function that $\{\boldsymbol{x}; g(\boldsymbol{x}, \tau + h) \le 0\} \to \{\boldsymbol{x}; g(\boldsymbol{x}, \tau) \le 0\}$ as $h \to 0$, by interchanging integration and partial differentiation, which is possible under the conditions above (see [62], p. 238) that

$$\lim_{h \to 0} \frac{1}{h} D_1(h) = \int\limits_{g(\boldsymbol{x}, \tau) \le 0} f_\tau(\boldsymbol{x}, \tau)\, d\boldsymbol{x}. \tag{2.73}$$

With equations (2.72) and (2.73) together the result is obtained. □

A problem of the last result is that the derivative is given as the sum of a surface and a domain integral. Since it is often difficult to calculate surface integrals, another form which transforms this surface integral into a domain integral might be useful. Uryas'ev has proved such a representation formula in [136], but the proof requires that the gradient $\nabla_{\boldsymbol{x}} g(\boldsymbol{x}, \tau)$ does not vanish in the domain $\{\boldsymbol{x}; g(\boldsymbol{x}, \tau) \le 0\}$. If the function $g(\boldsymbol{x}, \tau)$ is bounded from below and continuously differentiable with respect to $\boldsymbol{x}$, there must be a point $\boldsymbol{x}_0$ in the domain, where the gradient vanishes. Therefore this assumption is to restrictive. The following theorem shows that the result remains valid even if there are points where the gradient vanishes as long as the Hessian at these points is regular.

Theorem 22 *Given is a function $F(\tau)$ as in the theorem 21, p. 23 and the same conditions as in this theorem are fulfilled. Assume further that:*

1. *The function $g(\boldsymbol{x}, \tau)$ is twice continuously differentiable.*
2. *For a given $\tau \in I$ there is a finite number of points $\boldsymbol{y}_1, \ldots, \boldsymbol{y}_k$ in the interior of the domain $\{\boldsymbol{x}; g(\boldsymbol{x}, \tau) \leq 0\}$ with $\nabla_{\boldsymbol{y}} g(\boldsymbol{y}_i, \tau) = \mathbf{o}$ and the Hessian $\boldsymbol{H}_g(\boldsymbol{y}_i)$ is regular at all these points.*
3. *The dimension n is larger than two.*

Then the function $F(\tau)$ is differentiable at τ and we have for the derivative $F'(\tau)$ that

$$F'(\tau) = \int\limits_{g(\boldsymbol{x},\tau)\leq 0} \left[f_\tau(\boldsymbol{x}, \tau) - \operatorname{div}\left(\frac{f(\boldsymbol{x}, \tau) g_\tau(\boldsymbol{x}, \tau)}{|\nabla_{\boldsymbol{x}} g(\boldsymbol{x}, \tau)|^2} \nabla_{\boldsymbol{x}} g(\boldsymbol{x}, \tau) \right) \right] d\boldsymbol{x}. \tag{2.74}$$

PROOF: First we assume that the gradient does not vanish. Then the divergence theorem gives that

$$\int\limits_{g(\boldsymbol{x},\tau)\leq 0} \operatorname{div}\left(f(\boldsymbol{x}, \tau) g_\tau(\boldsymbol{x}, \tau) \frac{\nabla_{\boldsymbol{x}} g(\boldsymbol{x}, \tau)}{|\nabla_{\boldsymbol{x}} g(\boldsymbol{x}, \tau)|^2} \right) d\boldsymbol{x} \tag{2.75}$$

$$= \int\limits_{G(\tau)} f(\boldsymbol{y}, \tau) \frac{g_\tau(\boldsymbol{y}, \tau)}{|\nabla_{\boldsymbol{y}} g(\boldsymbol{y}, \tau)|} \langle \frac{\nabla_{\boldsymbol{y}} g(\boldsymbol{y}, \tau)}{|\nabla_{\boldsymbol{y}} g(\boldsymbol{y}, \tau)|}, \frac{\nabla_{\boldsymbol{y}} g(\boldsymbol{y}, \tau)}{|\nabla_{\boldsymbol{y}} g(\boldsymbol{y}, \tau)|} \rangle ds_\tau(\boldsymbol{y}).$$

Since the scalar product is always equal to unity, we get then

$$= \int\limits_{G(\tau)} f(\boldsymbol{y}, \tau) \frac{g_\tau(\boldsymbol{y}, \tau)}{|\nabla_{\boldsymbol{y}} g(\boldsymbol{y}, \tau)|} ds_\tau(\boldsymbol{y}). \tag{2.76}$$

But this is just the second term in equation (2.62), p. 23 in theorem 21 and so the result is correct in this case.

Now we assume that there is one point $\boldsymbol{y}_1$ in the interior of D where the gradient vanishes and the Hessian is regular at this point.

Then we can again use the divergence theorem for a slightly reshaped domain $F(\epsilon) = F \setminus \{\boldsymbol{x}; |\boldsymbol{x} - \boldsymbol{y}_1| \leq \epsilon\}$, where ϵ is chosen so small that $F(\epsilon) \subset F$. Then the gradient does not vanish througout $F(\epsilon)$ and we get using the result we just proved that

$$\int\limits_{F(\epsilon)} \operatorname{div}\left(f(\boldsymbol{y}, \tau) g_\tau(\boldsymbol{y}, \tau) \frac{\nabla_{\boldsymbol{y}} g(\boldsymbol{y}, \tau)}{|\nabla_{\boldsymbol{y}} g(\boldsymbol{y}, \tau)|^2} \right) d\boldsymbol{x} \tag{2.77}$$

$$= \underbrace{\int\limits_{G(\tau)} f(\boldsymbol{y}, \tau) \frac{g_\tau(\boldsymbol{y}, \tau)}{|\nabla_{\boldsymbol{y}} g(\boldsymbol{y}, \tau)|} ds_\tau(\boldsymbol{y})}_{=I_1(\epsilon)}$$

$$+\underbrace{\int_{K_\epsilon} \frac{f(\mathbf{y},\tau)g_\tau(\mathbf{y},\tau)}{|\nabla_{\mathbf{y}} g(\mathbf{y},\tau)|}\langle \frac{\mathbf{y}_1-\mathbf{y}}{|\mathbf{y}_1-\mathbf{y}|}, \frac{\nabla_{\mathbf{y}} g(\mathbf{y},\tau)}{|\nabla_{\mathbf{y}} g(\mathbf{y},\tau)|}\rangle \, ds_\epsilon(\mathbf{y})}_{=I_2(\epsilon)}.$$

Here K_ϵ is the sphere around $\mathbf{y}_1$ with radius ϵ and $ds_\epsilon(\mathbf{y})$ denotes surface integration over K_ϵ. For the second integral we have, since the absolute value of the scalar product is always less equal than unity, that

$$|I_2(\epsilon)| \leq K \int_{K_\epsilon} |\nabla_{\mathbf{y}} g(\mathbf{y},\tau)|^{-1} \, ds_\epsilon(\mathbf{y}) \tag{2.78}$$

with $K = \max_{\mathbf{y} \in F \setminus F(\epsilon)} |f(\mathbf{y},\tau)g_\tau(\mathbf{y},\tau)|$.

Making a Taylor expansion of the gradient at $\mathbf{y}_1$ gives

$$\nabla_{\mathbf{y}} g(\mathbf{y},\tau) = \mathbf{H}_g(\mathbf{y}_1)(\mathbf{y}-\mathbf{y}_1) + o(\epsilon). \tag{2.79}$$

Let $\lambda_1, \ldots, \lambda_n$ be the eigenvalues of the Hessian. Since $\mathbf{H}_g(\mathbf{y}_1)$ is regular, we get with $0 < \lambda_0 = \min\{|\lambda_1|, \ldots, |\lambda_n|\}$ that

$$|\nabla_{\mathbf{y}} g(\mathbf{y},\tau)| \geq \epsilon\lambda_0 + o(\epsilon). \tag{2.80}$$

We have then for the integral $I_2(\epsilon)$

$$|I_2(\epsilon)| \leq K \int_{K_\epsilon} [\epsilon\lambda_0 + o(\epsilon))]^{-1} \, ds_\epsilon(\mathbf{y}). \tag{2.81}$$

Evaluating this integral using equation (2.29), p. 16 gives then for $n > 2$ that

$$I_2(\epsilon) = o(\epsilon), \ \epsilon \to 0. \tag{2.82}$$

Here we need the assumption that $n > 2$. Therefore as $\epsilon \to 0$ the integral $I_2(\epsilon)$ tends to zero and we have in this case

$$\lim_{\epsilon\to 0} \int_{F(\epsilon)} \operatorname{div}\left(f(\mathbf{y},\tau)g_\tau(\mathbf{y},\tau)\frac{\nabla_{\mathbf{y}} g(\mathbf{y},\tau)}{|\nabla_{\mathbf{y}} g(\mathbf{y},\tau)|^2}\right) d\mathbf{x} \tag{2.83}$$

$$= \int_{G(\tau)} f(\mathbf{y},\tau)\frac{g_\tau(\mathbf{y},\tau)}{|\nabla_{\mathbf{y}} g(\mathbf{y},\tau)|} \, ds_\tau(\mathbf{y}).$$

This proves the result for the case of one stationary point. It is generalized in an analogous way for several points with vanishing gradients. □

Using the same arguments as in the theorem, we get a simple result, which shows how to transform a surface integral into a domain integral.

Corollary 23 *Given is a C^1-function $f : \mathbb{R}^n \to \mathbb{R}$ with $n > 2$ and a compact domain $F \subset \mathbb{R}^n$ defined by a C^2-function $g : \mathbb{R}^n \to \mathbb{R}$ by $F = \{\boldsymbol{x}; g(\boldsymbol{x}) \leq 0\}$. The gradient of g does not vanish on the surface $G = \{\boldsymbol{x}; g(\boldsymbol{x}) = 0\}$ and vanishes only at a finite number of points in the interior of F and at all these points the Hessian $\boldsymbol{H}_g(\boldsymbol{x})$ is regular. Then*

$$\int_G f(y)\, ds(y) = \int_F \operatorname{div}\left(f(\boldsymbol{x})\frac{\nabla g(\boldsymbol{x})}{|\nabla g(\boldsymbol{x})|}\right) d\boldsymbol{x}. \tag{2.84}$$

PROOF: As in the last theorem. □

2.6 Probability Distributions

2.6.1 Univariate distributions

A univariate random variable X is called a normally distributed random variable if its p.d.f. has the form

$$f(x) = \frac{1}{\sigma\sqrt{2\pi}} e^{-\frac{(x-\mu)^2}{2\sigma^2}}. \tag{2.85}$$

Here $\mathbb{E}(X) = \mu$ and $\operatorname{var}(X) = \sigma^2$. If $\mu = 0$ and $\sigma^2 = 1$ the distribution is called standard normal. The c.d.f. of this distributions is denoted by $\Phi(x)$, i.e.

$$\Phi(x) = \frac{1}{\sqrt{2\pi}} \int_{-\infty}^{x} e^{-y^2/2}\, dy. \tag{2.86}$$

Its density is denoted by $\varphi_1(x)$, i.e. $\varphi_1(x) = (2\pi)^{-1/2}\exp(-x^2/2)$.

For this c.d.f. there is the following asymptotic expansion (see [1], p. 932, 26.2.12.)

$$\Phi(-x) \sim \frac{\varphi_1(x)}{x}\left(1 + \sum_{n=1}^{\infty} \frac{(-1)^n(2n-1)!!}{x^{2n}}\right),\ x \to \infty. \tag{2.87}$$

Here the symbol $n!!$ denotes the product over all odd natural numbers less equal n. The expansion above is divergent. The error R_n for terminating the expansion at the n-th term

$$R_n = (-1)^{n+1}(2n+1)!! \int_x^{\infty} \frac{\varphi_1(y)}{y^{2n+2}}\, dy. \tag{2.88}$$

Often only the leading term in this expansion, Mill's ratio, is used

$$\Phi(-x) \sim \frac{\varphi_1(x)}{x},\ x \to \infty. \tag{2.89}$$

The square root of the sum of the squares of n independent random variables $X_1, \ldots, X_n$ with standard normal distributions $\sqrt{\sum_{i=1}^n X_i^2}$ has a χ_n-distribution. This distribution has the p.d.f.

$$\chi_n(x) = \frac{2x^{n-1}}{2^{n/2}\Gamma(n/2)} e^{-x^2/2} \text{ for } x \geq 0. \tag{2.90}$$

For the complementary c.d.f. of this distribution $\tilde{Q}(x|n) = \mathbb{P}(\sqrt{\sum_{i=1}^n X_i^2} > x)$ from equation 26.4.10, p. 941 in [1] the following asymptotic expansion is derived

$$\tilde{Q}(x|n) \sim \frac{2}{2^{n/2}\Gamma(n/2)} x^{n-2} e^{-x^2/2} = \frac{\chi_n(x)}{x}, \quad x \to \infty. \tag{2.91}$$

A random variable X has an exponential distribution with parameter λ, symbolically $\mathrm{Exp}(\lambda)$, if its p.d.f. has the form

$$f(x) = \lambda e^{-\lambda x} \text{ for } x \geq 0. \tag{2.92}$$

2.6.2 The n-dimensional normal distribution

Let $X_1, \ldots, X_n$ be an n-dimensional random vector with p.d.f.

$$f(\boldsymbol{x}) = (2\pi)^{-n/2} |\det(\boldsymbol{\Sigma})|^{-1/2} \exp\left(-\frac{1}{2}(\boldsymbol{x}-\boldsymbol{\mu})^T \boldsymbol{\Sigma}^{-1}(\boldsymbol{x}-\boldsymbol{\mu})\right). \tag{2.93}$$

Such a distribution is called an n-dimensional normal distribution. (We consider here only non-singular distributions.) Here $\boldsymbol{\mu} = (\mu_1, \ldots, \mu_n)$ is an n-dimensional vector and $\boldsymbol{\Sigma} = (\sigma_{ij})_{i,j=1,\ldots,n}$ is a positive definite $n \times n$-matrix and we have

$$\begin{aligned} \mathbb{E}(X_i) &= \mu_i \text{ for } i = 1, \ldots, n, & (2.94)\\ \mathrm{cov}(X_i, X_j) &= \sigma_{ij} \text{ for } i, j = 1, \ldots, n. & (2.95) \end{aligned}$$

Such a distribution with mean vector $\boldsymbol{\mu}$ and covariance matrix $\boldsymbol{\Sigma}$ is denoted by $\mathrm{N}_n(\boldsymbol{\mu}, \boldsymbol{\Sigma})$. The density of the $\mathrm{N}_n(\mathrm{o}, \boldsymbol{I}_n)$-distribution is denoted by $\varphi_n(\boldsymbol{x})$, i.e.

$$\varphi_n(\boldsymbol{x}) = (2\pi)^{-n/2} \exp\left(-\frac{1}{2}\sum_{i=1}^n x_i^2\right) = (2\pi)^{-n/2} \exp\left(-|\boldsymbol{x}|^2/2\right). \tag{2.96}$$

Let the random vector $\boldsymbol{X}$ consist of two vectors $\boldsymbol{X}_1 = (X_1, \ldots, X_r)$ and $\boldsymbol{X}_2 = (X_{r+1}, \ldots, X_n)$. The mean vector is split in the same way

$$\begin{aligned} \boldsymbol{\mu}_1 = (\mu_1, \ldots, \mu_r) &= (\mathbb{E}(X_1), \ldots, \mathbb{E}(X_r)) & (2.97)\\ \boldsymbol{\mu}_2 = (\mu_{r+1}, \ldots, \mu_n) &= (\mathbb{E}(X_{r+1}), \ldots, \mathbb{E}(X_n)). \end{aligned}$$

Then the covariance matrix has the form

$$\boldsymbol{\Sigma} = \begin{pmatrix} \boldsymbol{\Sigma}_{11} & \boldsymbol{\Sigma}_{12} \\ \boldsymbol{\Sigma}_{21} & \boldsymbol{\Sigma}_{22} \end{pmatrix}. \tag{2.98}$$

Here $\boldsymbol{\Sigma}_{11}$ is the covariance matrix of the random vector $\boldsymbol{X}_1$, $\boldsymbol{\Sigma}_{22}$ is the covariance matrix of the random vector $\boldsymbol{X}_2$ and $\boldsymbol{\Sigma}_{12}$ is the matrix of the covariances between the components of the random vectors $\boldsymbol{X}_1$ and $\boldsymbol{X}_2$.

Theorem 24 *The conditional distribution of $\boldsymbol{X}_2$ given $\boldsymbol{X}_1$ is an $(n-r)$-dimensional normal distribution*

$$\mathrm{N}_{n-r}(\boldsymbol{\mu}_2 + \boldsymbol{\Sigma}_{21}\boldsymbol{\Sigma}_{11}^{-1}(\boldsymbol{X}_1 - \boldsymbol{\mu}_1), \boldsymbol{\Sigma}_{22} - \boldsymbol{\Sigma}_{21}\boldsymbol{\Sigma}_{11}^{-1}\boldsymbol{\Sigma}_{12}), \tag{2.99}$$

where $\boldsymbol{\Sigma}_{11}^{-1}$ is the inverse of the matrix $\boldsymbol{\Sigma}_{11}$.

PROOF: See [111], chapter 8, p. 552. □

2.6.3 Calculation of normal integrals

Here some integrals are evaluated, which appear in the following. More detailed discussions and results can be found in [100] and [106].

Lemma 25

$$\int_{-\infty}^{\infty} |x|\varphi_1(x)\,dx = \sqrt{\frac{2}{\pi}}. \tag{2.100}$$

PROOF:

$$\int_{-\infty}^{\infty} |x|\varphi_1(x)\,dx = 2\int_0^{\infty} x\varphi_1(x)\,dx \tag{2.101}$$

$$= 2\int_0^{\infty}\left[-\frac{d\varphi_1(x)}{dx}\right]dx = 2\cdot\varphi_1(0) = 2\frac{1}{\sqrt{2\pi}} = \sqrt{\frac{2}{\pi}}.$$

□

Lemma 26 *Given are a negative definite $n\times n$-matrix $\boldsymbol{A}$ and two n-dimensional vectors $\boldsymbol{b}$ and $\boldsymbol{c}$. Then*

$$\int_{\mathbb{R}^n} \exp(\frac{1}{2}\boldsymbol{x}^T\boldsymbol{A}\boldsymbol{x})\,d\boldsymbol{x} = \frac{(2\pi)^{n/2}}{\sqrt{|\det(\boldsymbol{A})|}}, \tag{2.102}$$

$$\int_{\mathbb{R}^n} \exp(\boldsymbol{b}^T\boldsymbol{x} + \frac{1}{2}\boldsymbol{x}^T\boldsymbol{A}\boldsymbol{x})\,d\boldsymbol{x} = \frac{(2\pi)^{n/2}}{\sqrt{|\det(\boldsymbol{A})|}}\exp(\frac{1}{2}\boldsymbol{b}^T\boldsymbol{A}^{-1}\boldsymbol{b}), \tag{2.103}$$

$$\int_{\mathbb{R}^n} |\boldsymbol{c}^T\boldsymbol{x}|\exp(\frac{1}{2}\boldsymbol{x}^T\boldsymbol{A}\boldsymbol{x})\,d\boldsymbol{x} = 2(2\pi)^{(n-1)/2}\left|\frac{\boldsymbol{c}^T\boldsymbol{A}^{-1}\boldsymbol{c}}{\det(\boldsymbol{A})}\right|^{1/2}. \tag{2.104}$$

PROOF: The first two assertions are shown in [100], p. 19.

To prove the last assertion, we assume that $-\boldsymbol{A}$ is the inverse of a covariance matrix of an n-dimensional normal distribution $\mathrm{N}(\mathbf{o}, -\boldsymbol{A}^{-1})$. The random vector with this distribution is denoted by $\boldsymbol{X}$. Then the coordinates are rotated by multiplying by an orthogonal matrix $\boldsymbol{T} = (\boldsymbol{t}_1, \ldots, \boldsymbol{t}_n)^T$ with $\boldsymbol{t}_1 = |\boldsymbol{c}|^{-1}\boldsymbol{c}$ and then a new random vector $\boldsymbol{Y}$ is defined by $\boldsymbol{Y} = \boldsymbol{T}\boldsymbol{X}$. Then we have

$$\frac{\sqrt{|\det(\boldsymbol{A})|}}{(2\pi)^{n/2}} \int_{\mathbb{R}^n} |\boldsymbol{c}^T\boldsymbol{x}| \exp(\frac{1}{2}\boldsymbol{x}^T\boldsymbol{A}\boldsymbol{x})\, d\boldsymbol{x} \tag{2.105}$$

$$= |\boldsymbol{c}| \left[\frac{\sqrt{|\det(\boldsymbol{A})|}}{(2\pi)^{n/2}} \int_{\mathbb{R}^n} |y_1| \exp(\frac{1}{2}\boldsymbol{y}^T\boldsymbol{T}^T\boldsymbol{A}\boldsymbol{T}\boldsymbol{y})\, d\boldsymbol{y}\right],$$

where the term in the brackets is the first absolute moment of a normally distributed random variable Y_1 with mean zero and variance $|\boldsymbol{c}|^{-2}\boldsymbol{c}^T\boldsymbol{A}^{-1}\boldsymbol{c}$; so we get for the integral

$$= |\boldsymbol{c}|\sqrt{\frac{2}{\pi}\mathrm{var}(Y_1)} = \sqrt{\frac{2}{\pi}|\boldsymbol{c}^T\boldsymbol{A}^{-1}\boldsymbol{c}|}. \tag{2.106}$$

Multiplying by $(2\pi)^{n/2}|\det(\boldsymbol{A})|^{-1/2}$ the result is obtained. □

2.7 Convergence of Probability Distributions

Definition 10 *A sequence F_n of k-dimensional distribution functions is called weakly convergent to a distribution function F iff*

$$\lim_{n\to\infty} F_n(\boldsymbol{x}) = F(\boldsymbol{x}) \tag{2.107}$$

for all points $\boldsymbol{x} \in \mathbb{R}^k$ at which F is continuous. Symbolically we write then $F_n \to F$.

Definition 11 *If $\boldsymbol{X}_n$ and $\boldsymbol{X}$ are random vectors with distribution functions F_n and F, then the sequence $\boldsymbol{X}_n$ is called to be convergent in distribution to $\boldsymbol{X}$ iff $F_n \to F$. Symbolically we write*

$$\boldsymbol{X}_n \xrightarrow{\mathrm{D}} \boldsymbol{X}. \tag{2.108}$$

Theorem 27 (Cramer-Wold-Device) *A sequence $(\boldsymbol{X}_n)_{n\in\mathbb{N}}$ of k-dimensional random vectors $\boldsymbol{X}_n = (X_{n1}, \ldots, X_{nk})$ converges in distribution to a random vector $\boldsymbol{X} = (X_1, \ldots, X_k)$ iff for all $(t_1, \ldots, t_k) \in \mathbb{R}^k$ $\sum_{j=1}^k t_j X_{nj} \xrightarrow{\mathrm{D}} \sum_{j=1}^k t_j X_j$.*

PROOF: A proof is given in [13], p. 335. □

Using this device the convergence of n-dimensional random vectors can be proved by showing the convergence of linear combinations of components of these vectors.

Definition 12 *Let X be a random variable. The moment generating function of X is defined by*

$$M(s) = \mathbb{E}(e^{sX}) = \int_{-\infty}^{\infty} e^{sx}\, dF(x) \tag{2.109}$$

for all $s \in \mathbb{R}$ with $M(s) < \infty$.

If the c.d.f. $F(x)$ has a p.d.f. $f(x)$, $M(s)$ can be written as

$$M(s) = \int_{-\infty}^{\infty} e^{sx} f(x)\, dx\ . \tag{2.110}$$

The standard normal distribution $\mathrm{N}(0,1)$ has the moment generating function (see [13], p. 242)

$$M(s) = e^{s^2/2}\ . \tag{2.111}$$

For an arbitrary normal distribution $\mathrm{N}(\mu, \sigma^2)$ instead we have

$$M(s) = e^{(s\sigma)^2/2 + s\mu}\ . \tag{2.112}$$

An exponential distribution $\mathrm{Exp}(\lambda)$ has the moment generating function

$$M(s) = \frac{\lambda}{\lambda - s}\ . \tag{2.113}$$

The convergence of random variables can be shown by proving the convergence of the corresponding moment generating functions.

Theorem 28 *Assume that the random variables X_n $(n \in \mathbb{N})$ and X have moment generating functions $M_n(s)$ and $M(s)$ defined in the interval $[-\delta, \delta]$ with $\delta > 0$ such that*

$$M_n(s) \to M(s)$$

for each $s \in [-\delta, \delta\ ]$, then

$$X_n \xrightarrow{D} X, \tag{2.114}$$

i.e. the random variables X_n converge in distribution towards X.

PROOF: See [13], p. 345. □

By combining the last two theorems a convergence theorems for multivariate random vectors can be derived.

Theorem 29 *A sequence* $(\boldsymbol{X}_n)_{n\in\mathbb{N}}$ *of k-dimensional random vectors* $\boldsymbol{X}_n = (X_{n1},\ldots,X_{nk})$ *converges in distribution to a random vector* $\boldsymbol{X} = (X_1,\ldots,X_k)$ *iff for all* $\boldsymbol{t} = (t_1,\ldots,t_k) \in \mathbb{R}^k$ *the moment generating functions* $M_{n,\boldsymbol{t}}(s)$ *of the random variables* $\sum_{j=1}^k t_j X_{nj}$ *converge towards the moment generating function* $M_{\boldsymbol{t}}(s)$ *of the random variable* $\sum_{j=1}^k t_j X_j$ *for any s in an interval* $[-\delta_{\boldsymbol{t}}, \delta_{\boldsymbol{t}}]$ *with* $\delta_{\boldsymbol{t}} > 0$.

PROOF: From theorem 27, p. 31 follows that from the convergence of all linear combinations of the components $X_{n1},\ldots,X_{nk}$, i.e. of the random variables $\sum_{i=1}^k t_i X_{ni}$ to the random variable $\sum_{i=1}^k t_i X_i$ follows the convergence of the random vectors $\boldsymbol{X}_n$ to $\boldsymbol{X}$. But the convergence of such a linear combination can be shown by proving that its moment generating function converges to the moment generating function of $\sum_{i=1}^k t_i X_i$. □

Chapter 3

Asymptotic Analysis

3.1 The Topic of Asymptotic Analysis

The main topic of asymptotic analysis is the study of the behavior of functions when the argument approaches certain limits on the boundary of the domain of definition of the function. In a more abstract setting, for two metric spaces X and Y and a function $f : X' \to Y$, where $X' \subset X$ with $X' \neq X$, asymptotic analysis studies the behavior of the function $f(x)$, as $x \to x_0$, where x_0 is a limit point of X' with $x_0 \notin X'$. Certainly it is also possible to study a function in the neighborhood of points inside its domain of definition with the methods of asymptotic analysis, but the interesting and important results are concerned with boundary points.

One of the basic ideas is to find a "nice" function $g(x)$ such that as $x \to x_0$ we have

$$f(x) = g(x) + r(x) \tag{3.1}$$

with $r(x)$ being "small" in comparison with $g(x)$ for x "near" x_0. "Nice " means here that $g(x)$ should be of a simpler form as $f(x)$, but describing the structure of $f(x)$ near x_0 "sufficiently good".

This is different from the problems of classical analysis where in general the behavior of f as $x \to x_0$ with $x \in X'$ is studied. If f is continuous at x_0, we have $\lim_{x \to x_0} f(x) = f(x_0)$. If f is differentiable at x_0, the function can be approximated still more precisely by its Taylor expansion at this point. For a function $f(x_1, \ldots, x_n)$, which is differentiable at a point $(x_1^0, \ldots, x_n^0)$, we can use for example the first order Taylor expansion to describe the function near this point

$$f(x_1, \ldots, x_n) \approx f(x_1^0, \ldots, x_n^0) + \sum_{i=1}^{n} \left. \frac{\partial f(x)}{\partial x_i} \right|_{\boldsymbol{x}=\boldsymbol{x}_0} (x_i - x_i^0). \tag{3.2}$$

If this is not sufficient, for example, if we are looking for extrema of the function, we can make higher order Taylor expansions. Finally, an analytic C^∞-function is given by its Taylor expansion at a point.

What makes asymptotic analysis a little bit tricky is the fact that here we are interested in the behavior of the function near a point where the function is not defined. Therefore here a more pathological behavior is possible near these points on the boundary.

EXAMPLE: Consider the function

$$f(z) = \exp(-1/z) \tag{3.3}$$

defined in the complex plane with the exception of the origin. In the region $\{z; \mathrm{Re}(z) > 0\}$ the function approaches zero as $z \to 0$, but for in the region $\{z; \mathrm{Re}(z) \leq 0\}$ the function shows a chaotic behavior as $z \to 0$, since 0 is an essential singularity of this function. □

Standard problems for asymptotic approximations are:

- The evaluation of integrals
- The computation of solutions of equations
- The evaluation of sums

Why use asymptotic methods in cases, where numerical solutions are available ? If there is only one number needed, it may be certainly easier to calculate it by maybe crude numerical methods. But in general, asymptotic expansions give much more insight in the structure of the problem and how various parameters influence the result. Deriving such results is surely more complicated, but we gain much more information by doing it this way. For example, using advanced integration methods it is possible nowadays to compute the distribution function of sums of random variables very accurately, but a result as the central limit theorem gives much more understanding what happens in the limit. Secondly sometimes even now the numerical difficulties are such that an asymptotic result is better. Thirdly, asymptotic approximations are a good starting point for further numerical or Monte Carlo methods.

Textbooks about asymptotic analysis are [127], [105], [15], [59] and [140]. The first book by Sirovich gives a broad overview about asymptotic methods, but is quite sloppy about proofs. The book of Olver [105] covers only univariate functions, but in a very detailed form. The book of Bleistein and Handelsman [15] treats the multivariate case in chapter 8. Wong [140] considers bivariate integrals in chapter VIII and multivariate integrals in chapter IX.

3.2 The Comparison of Functions

The most important cases in asymptotic analysis are when the function $f(x)$ approaches zero or infinity as $x \to x_0$. But to describe this behavior more precisely, we need some ideas to measure the velocity towards infinity or zero. For example, all the following functions approach infinity as $x \to \infty$

$$\ln(\ln(x)), \ln(x), \sqrt{x}, x, x^2, 2^x, \exp(x), \exp(\exp(x)), \tag{3.4}$$

but it can be seen easily, using l'Hospital's rule, that from the left to the right the velocity of convergence towards infinity increases. To find simple functions which model the asymptotic form of more complicated functions we have to introduce order relations between functions.

To compare the behavior of two functions, often the following symbols o and O, introduced by Landau, are used.

Definition 13 *A function $f : M \to \mathbb{R}$ is of order O of the function $g : M \to \mathbb{R}$ as $x \to x_0$ with $x \in M$ if there is a constant K and a neighborhood U_K of x_0 such that*

$$| f(x) | \leq K | g(x) | \quad \text{for all} \quad x \in U_K \cap M. \tag{3.5}$$

Then we say that as $x \to x_0$, f is large "O" of g and write symbolically

$$f(x) = O(g(x)), \ x \to x_0. \tag{3.6}$$

Definition 14 *A function $f : M \to \mathbb{R}$ is of order o of the function $g : M \to \mathbb{R}$ as $x \to x_0$ with $x \in M$ if for all constants $K > 0$ there is a neighborhood U_K of x_0 such that*

$$| f(x) | \leq K | g(x) | \quad \text{for all} \quad x \in U_K \cap M. \tag{3.7}$$

We write symbolically

$$f(x) = o(g(x)), \ x \to x_0. \tag{3.8}$$

and say that, as $x \to x_0$, f is small "o" of g.

For these order relations there are a number of useful relations following from the definitions above. We have:

1. $O(O(f)) = O(f)$,
2. $O(o(f)) = o(f)$,
3. $o(O(f)) = o(f)$,
4. $o(o(f)) = o(f)$,
5. $O(fg) = O(f)O(g)$,
6. $O(f) \cdot o(g) = o(fg)$,
7. $O(f) + O(f) = O(f)$,
8. $o(f) + o(f) = o(f)$,
9. $o(f) + O(f) = O(f)$.

The meaning of the first is for example that if $g = O(h)$ and $h = O(f)$, as $x \to x_0$, then $g = O(f)$.

A stronger asymptotic relation between functions is the following.

Definition 15 *If for two functions $f : M \to \mathbb{R}$ and $g : M \to \mathbb{R}$ (both non vanishing in a neighborhood of x_0) as $x \to x_0$ with $x \in M$*

$$\lim_{x \to x_0} \frac{f(x)}{g(x)} = 1, \tag{3.9}$$

these functions are called asymptotically equivalent and we write symbolically

$$f(x) \sim g(x), \; x \to x_0. \tag{3.10}$$

This is an equivalence relation between functions. It gives more information than the "O"- and "o"- relations. The relation $f(x) \sim g(x)$ as $x \to x_0$ means that the relative error of approximating $f(x)$ by $g(x)$ converges to zero as $x \to x_0$, since

$$\lim_{x \to x_)} \frac{g(x) - f(x)}{g(x)} = 1 - \lim_{x \to x_0} \frac{f(x)}{g(x)} = 0. \tag{3.11}$$

Another way of expressing $f(x) \sim g(x)$ as $x \to x_0$ is the equation

$$f(x) = g(x) + o(f(x)). \tag{3.12}$$

In many cases the main interest is to find for a given function $f(x)$ an asymptotically equivalent function $g(x)$ which is of a simpler form.

Asymptotic equations can be multiplied, divided and raised to arbitrary powers. If $f \sim \phi$ and $g \sim \psi$ as $x \to x_0$, we have

1. $fg \sim \phi\psi$,
2. $f/g \sim \phi/\psi$,
3. $f^\alpha \sim \phi^\alpha$.

Asymptotic equations can be added only if all summands are positive (or negative). The following example shows that elsewhere wrong results are obtained.

EXAMPLE:

$$x - x^2 \sim 3x - x^2, \; x \to \infty, \tag{3.13}$$
$$x^2 \sim x^2, \; x \to \infty. \tag{3.14}$$

By adding these asymptotic equations we would get that x is asymptotically equivalent to $3x$ as $x \to \infty$, which is wrong. □

Consider now that $f(x) \sim g(x)$ as $x \to x_0$. Under which conditions then follows that $h(f(x)) \sim h(g(x))$ as $x \to x_0$ for a function h?

Theorem 30 *Let be given two functions $f(x)$ and $g(x)$ on a set M with*

$$f(x) \sim g(x), \ x \to x_0. \tag{3.15}$$

Further is given a function h on a set D such that $f(x) \in D$ and $g(x) \in D$ for all $x \in M$. If there is a closed set $X \subset D$ (which may contain infinity) with $f(x) \in X$ and $g(x) \in X$ for all $x \in M$ and such that h is continuous on X and $h(x) \neq 0$ for all $x \in X$, then

$$h(f(x)) \sim h(g(x)), \ x \to x_0. \tag{3.16}$$

PROOF: (See [11], p. 14) If $h(f(x)) \not\sim h(g(x))$ as $x \to x_0$, there must be a $\delta > 0$ and a sequence $x_n \to x_0$ with $| h(f(x_n))/h(g(x_n)) - 1 | \geq \delta$ for all $n \in \mathbb{N}$. The sequence $f(x_n)$ has (at least) one limit point z in D. Let now be x_{n_k} a subsequence of x_n with $f(x_{n_k}) \to z$, then due to equation (3.15) $g(x_{n_k}) \to z$. Since $h(x)$ is continuous, we get therefore $h(f(x_{n_k})) \to h(z)$ and $h(g(x_{n_k})) \to h(z)$. From this follows using $h(z) \neq 0$ that $h(f(x_{n_k}))/h(g(x_{n_k})) \to 1$. But this contradicts the assumptions about the sequence. □

It is important to notice for which cases the theorem above is not valid. If we have a function f with $f(x) \to 0$ as $x \to \infty$, then we can conclude that

$$f(x) \sim g(x), \ x \to \infty \Rightarrow \ln(f(x)) \sim \ln(g(x)), \ x \to \infty. \tag{3.17}$$

But in general we have

$$\ln(f(x)) \sim \ln(g(x)), \ x \to \infty \not\Rightarrow f(x) \sim g(x), \ x \to \infty. \tag{3.18}$$

Therefore it is often easier to derive an asymptotic approximation for the logarithm of a function than for the function itself. But the approximation of the logarithm gives less information as shown in the next example.

EXAMPLE: Given are the functions $f(x) = x^n \cdot e^{-x}$ und $g(x) = e^{-x}$. Then

$$\lim_{x \to \infty} \frac{\ln(f(x))}{\ln(g(x))} = \lim_{x \to \infty} \frac{n \cdot \ln(x) - x}{-x} = 1, \tag{3.19}$$

but for the functions themselves we get

$$\lim_{x \to \infty} \frac{f(x)}{g(x)} = \lim_{x \to \infty} \frac{x^n \cdot e^{-x}}{e^{-x}} = \lim_{x \to \infty} x^n = \infty. \tag{3.20}$$

□

The differentiation of order relations and asymptotic equations is in general possible only under some restrictions. Results are given in [105], p. 8-11 for holomorphic functions and functions with monotone derivatives. A more complete review of these results can be found in [11]. The integration is much easier as shown in the following lemma.

Lemma 31 *If f and g are continuous functions on (a, ∞) with $g > 0$, we have for integrals the following results*

1. *If $\int_a^\infty g(x)dx = \infty$, then*

   (a) $f = O(g), x \to \infty \;\Rightarrow\; \int_a^x f(y)dy = O(\int_a^x g(y)dy), x \to \infty,$

   (b) $f = o(g), x \to \infty \;\Rightarrow\; \int_a^x f(y)dy = o(\int_a^x g(y)dy), x \to \infty,$

   (c) $f \sim c \cdot g$ *with* $c \neq 0$, $x \to \infty \;\Rightarrow\; \int_a^x f(y)dy \sim c \cdot \int_a^x g(y)dy, x \to \infty.$

2. *If $\int_a^\infty g(x)dx < \infty$, then*

   (a) $f = O(g), x \to \infty \;\Rightarrow\; \int_x^\infty f(y)dy = O(\int_x^\infty g(y)dy), x \to \infty,$

   (b) $f = o(g), x \to \infty \;\Rightarrow\; \int_x^\infty f(y)dy = o(\int_x^\infty g(y)dy), x \to \infty,$

   (c) $f \sim c \cdot g$ *with* $c \neq 0$, $x \to \infty \;\Rightarrow\; \int_x^\infty f(y)dy \sim c \cdot \int_x^\infty g(y)dy, x \to \infty.$

PROOF: We prove the first three statements. If $f = O(g)$ as $x \to \infty$, we can find using the fact that f and g are continuous and $g > 0$ a constant K such $|f(y)| \leq Kg(y)$ for all $y \geq a$ and not only for a neighborhood of ∞ and therefore

$$\left|\int_a^x f(y)dy\right| \leq K \cdot \int_a^x g(y)dy. \tag{3.21}$$

In the second case there exists for all $\epsilon > 0$ an $x_\epsilon \geq a$ such that $|f(y)| \leq \epsilon g(y)$ for all $y \geq x_\epsilon$ and hence for all $x \geq x_\epsilon$

$$\left|\int_{x_\epsilon}^x f(y)dy\right| \leq \epsilon \int_{x_\epsilon}^x g(y)dy \leq \epsilon \int_a^x g(y)dy. \tag{3.22}$$

Since the integral $\int_a^\infty g(y)dy$ is divergent, we can choose an x so large that

$$\epsilon \int_a^x g(y)dy \geq \left|\int_a^{x_\epsilon} f(y)dy\right|. \tag{3.23}$$

The last two relations together give

$$\left|\int_a^x f(y)dy\right| \leq 2\epsilon \int_a^x g(y)dy. \tag{3.24}$$

This completes the proof. To prove the third statement we use the last result for the function $f - cg$ which is $o(g)$, if $f \sim cg$. □

3.3 Asymptotic Power Series and Scales

To describe the asymptotic behavior of a function more precisely, one can define general expansions.

Definition 16 *Let f be a continuous function on the set $M \subset \mathbb{R}$ with x_0 a limit point of this set. The formal power series $\sum_{n=0}^{\infty} a_n(x-x_0)^n$ is said to be an asymptotic power series expansion of f, as $x \to x_0$, in M if the following equations*

$$\lim_{x\to x_0}\left\{(x-x_0)^{-m}\left[f(x)-\sum_{n=0}^{m} a_n(x-x_0)^n\right]\right\}=0 \tag{3.25}$$

hold for all $m \in \mathbb{N}$ or in an equivalent formulation

$$f(x)-\sum_{n=0}^{m} a_n(x-x_0)^n = O\left((x-x_0)^{m+1}\right). \tag{3.26}$$

We write symbolically

$$f(x)\sim\sum_{n=0}^{\infty} a_n(x-x_0)^n,\ x\to x_0. \tag{3.27}$$

Definition 17 *If the equation (3.25) is satisfied only for $m = 0,\ldots,N-1$, but not for $m \geq N$, we say that $\sum_{n=0}^{N-1} a_n(x-x_0)^n$ is an asymptotic power series of f as $x \to x_0$ to N terms and write*

$$f(x)\sim\sum_{i=0}^{N-1} a_i(x-x_0)^i,\ x\to x_0. \tag{3.28}$$

Definition 18 *The formal series $\sum_{n=0}^{\infty} a_n x^{-n}$ is said to be an asymptotic power series expansion of f as $x \to \infty$ if for all $m \in \mathbb{N}$ always*

$$\lim_{x\to\infty}\left\{x^m\left[f(x)-\sum_{n=0}^{m} a_n x^{-n}\right]\right\}=0. \tag{3.29}$$

An equivalent formulation is

$$f(x)=\sum_{n=0}^{m} a_n x^{-n}+O\left(x^{-(m+1)}\right),\ x\to\infty. \tag{3.30}$$

for all $m \in \mathbb{N}$. If these conditions are fulfilled, we write

$$f(x)\sim\sum_{n=0}^{\infty} a_n x^{-n},\ x\to\infty. \tag{3.31}$$

In the same way as above we define an asymptotic power series expansion of f as $x \to \infty$ to N terms if the conditions above are satisfied only for $m = 0, \ldots, N-1$. If there exists an asymptotic power series expansion of a function, it is unique (see [15], p. 16).

In many cases asymptotic expansions of functions have to be made with respect to more complicated functions, which are no power series.

Definition 19 *A sequence of functions $\{\phi_n(x)\}$ with $n \in I\!N$ is called an asymptotic scale or asymptotic sequence as $x \to x_0$ in M if for all n the function $\phi_n(x)$ is a continuous function on M and further*

$$\phi_{n+1}(x) = o(\phi_n(x)), \quad x \to x_0. \tag{3.32}$$

In [15] the term "asymptotic sequence" is used. Here in the following we use "asymptotic scale" as in [11].

EXAMPLE: The following sequences of functions are asymptotic scales:

1. $\{(x - x_0)^{\gamma_n}\}$, $x \to x_0$, $\gamma_n \in \mathbb{C}$, $\mathrm{Re}(\gamma_{n+1}) > \mathrm{Re}(\gamma_n)$,
2. $\{x^{-\gamma_n}\}$, $x \to \infty$, $\gamma_n \in \mathbb{C}$, $\mathrm{Re}(\gamma_{n+1}) > \mathrm{Re}(\gamma_n)$,
3. $\{[g(x)]^n\}$, $x \to x_0$, $g(x_0) = 0$,
4. $\{g(x)\phi_n(x)\}, x \to x_0$.

Here in 4. the functions $\{\phi_n\}$ form an asymptotic scale as $x \to x_0$, while $g(x)$ in 3. and 4. is continuous and not identically zero in any neighborhood of x_0. □

Now we can define an expansion of a function with respect to such a scale.

Definition 20 *Let f be a continuous function defined on $M \subset I\!R$ and let the sequence $\{\phi_n(x)\}$ be an asymptotic scale as $x \to x_0$ in M. Then the formal series $\sum_{n=1}^{\infty} a_n \phi_n(x)$ is said to be an asymptotic expansion of $f(x)$ as $x \to x_0$ with respect to $\{\phi_n(x)\}$ if for all $m \in I\!N$ always*

$$\lim_{x \to x_0} \frac{f(x) - \sum_{n=1}^{m} a_n \phi_n(x)}{\phi_m(x)} = 0, \tag{3.33}$$

or in an equivalent form

$$f(x) = \sum_{n=1}^{m} a_n \phi_n(x) + O\left(\phi_{m+1}(x)\right). \tag{3.34}$$

Symbolically

$$f(x) \sim \sum_{n=1}^{\infty} a_n \phi_n(x), \quad x \to x_0. \tag{3.35}$$

Definition 21 *If the equation (3.33) holds only for $m = 1, \ldots, N$, then we write*

$$f(x) \sim \sum_{n=1}^{N} a_n \phi_n(x), \quad x \to x_0. \tag{3.36}$$

This is called an asymptotic expansion of f to N terms with respect to the asymptotic scale $\{\phi_n(x)\}$ or an expansion of Poincaré type.

If a function has an asymptotic expansion with respect to an asymptotic scale, this expansion is unique and its coefficients are given successively by

$$a_m = \lim_{x \to x_0} \left[f(x) - \sum_{i=1}^{m-1} a_i \phi_i(x) \right] / \phi_m(x). \tag{3.37}$$

This is shown in the next theorem.

Theorem 32 *Let the function f have an asymptotic expansion to N terms with respect to the asypmtotic scale $\{\phi_n\}$ as $x \to x_0$ in the form*

$$f(x) \sim \sum_{n=1}^{N} a_n \phi_n(x), \quad x \to x_0. \tag{3.38}$$

Then the coefficients $a_1, \ldots, a_n$ are uniquely determined.

PROOF: See [15], p. 16-17. □

EXAMPLE: But it is possible that different functions have the same asymptotic expansion. For example the functions

$$\frac{1}{x+1} \text{ and } \frac{1}{x+1} + \exp(-x) \tag{3.39}$$

both have the asymptotic expansion with respect to the scale $\{x^{-n}\}$, namely

$$\sum_{n=1}^{\infty} \frac{(-1)^{n-1}}{x^n} \tag{3.40}$$

as $x \to \infty$, since $x^n \exp(-x)$ tends to zero as $x \to \infty$ for all $n \in \mathbb{N}$. □

The following two theorems about integration and differentiation are from [15]. The integration of asymptotic relations is possible under weak conditions.

Theorem 33 *Given is an asymptotic scale $\{\psi_n(x)\}$ of positive functions on $I = (a, x_0)$ as $x \to x_0$ such that for all $x \in (a, x_0)$ the integrals*

$$\Psi_n(x) = \int_x^{x_0} |\psi_n(y)|\, dy \tag{3.41}$$

exist. Then the $\Psi_n(x)$ form again an asymptotic scale. If f is a function such that as $x \to x_0$

$$f(x) \sim \sum_{n=1}^{N} a_n \psi_n(x), \; x \to x_0 \tag{3.42}$$

and if further the function

$$g(x) = \int_x^{x_0} f(t)\, dt$$

exists for all $x \in (b, x_0)$ with $a \leq b < x_0$, then $g(x)$ has an asymptotic expansion to N terms as $x \to x_0$ with respect to the asymptotic scale $\{\Psi_n(x)\}$ in the form

$$g(x) \sim \sum_{n=1}^{N} a_n \Psi_n(x), \; x \to x_0. \tag{3.43}$$

PROOF: See [15], p. 30-31. □

This shows that integrating an asymptotic expansion with respect to an asymptotic scale with positive functions gives immediately the asymptotic expansion with respect to the integrated scale functions. Differentiating them is not so easy, we must assume that an expansion of the derivative exists.

Theorem 34 *Given is an asymptotic scale $\{\psi_n(x)\}$ as $x \to x_0$ on an interval $I = (a, x_0)$. Assume that $\psi_1(x) = o(1)$ as $x \to x_0$ and that all $\psi_n(x)$ are differentiable in I with derivative functions $\chi_n(x) = \psi_n'(x)$ such that these functions are again an asymptotic scale $\{\chi_n(x)\}$ as $x \to x_0$ in I. A differentiable function $f(x)$ in I has an asymptotic expansion to N terms with respect to the asymptotic scale $\{\psi(x)\}_n$*

$$f(x) - f(x_0) \sim \sum_{n=1}^{N} a_n \psi_n(x), \; x \to x_0. \tag{3.44}$$

If then
a) the function $\chi_n(x)$ is positive in an interval (b, x_0) with $a \leq b < x_0$ and
b) the function $f'(x)$ has an asymptotic expansion to N terms with respect to $\{\chi_n(x)\}$,
this expansion has the form

$$f'(x) \sim \sum_{n=1}^{N} a_n \chi_n(x), \; x \to x_0. \tag{3.45}$$

PROOF: See [15], p. 32. □

It may occur that the derivative $f'(x)$ of a function $f(x)$ does not have an asymptotic expansion with respect to an asymptotic scale $\{\phi_n\}$, even in the case that the function $f(x)$ itself has an expansion with respect to the asymptotic scale $\{\Phi_n\}$ with $\phi_n = \Phi'_n$, see [15], p. 31.

3.4 Deriving Asymptotic Expansions

The derivation of an asymptotic approximation or expansion for a function $f(x)$ as $x \to x_0$ can not be made schematically. If we consider for example the function $f(x) = e^{-x}$ as $x \to \infty$ and the asymtotic scale $\{x^{-n}\}$, then an asymptotic expansion of $f(x)$ with respect to this scale would be

$$f(x) \sim \sum_{i=1}^{\infty} 0 \cdot x^{-i}, \; x \to \infty. \tag{3.46}$$

This says only that $f(x)$ approaches zero faster than all functions of the asymptotic scale as $x \to \infty$ and gives no further information about the asymptotic behavior of $f(x)$. Therefore before making an expansion we have to choose a suitable scale. This is done usually by making some rough prior estimates of the order of magnitude of the function under consideration.

There are normally three (or two) steps in deriving an asymptotic expansion for a function $f(x)$ as $x \to x_0$:

1. Choose an asymptotic scale $\{\psi_n(x)\}$,
2. Calculate the expansion coefficients a_n for the asymptotic relation
$$f(x) \sim \sum_{n=1}^{\infty} a_n \psi_n(x), \; x \to x_0, \tag{3.47}$$
3. Calculate (if possible) error estimates for the approximation errors
$$f(x) - \sum_{n=1}^{N} a_n \psi_n(x) \tag{3.48}$$
for $N = 1, 2, \ldots$.

Often there is no possibility to find the error estimates in step 3. Then judgements about the quality of the approximations can be based only on numerical examples, which give some idea about the magnitude of the error. This happens especially in the multivariate case.

An important new method for calculating approximations is the combination of analytic and Monte Carlo methods. Here the analytic approximations found by asymptotic analysis are used as an initial estimate of the integral and then this estimate is improved by Monte Carlo importance sampling methods which use the information about the structure of the integrand found by the analytic method.

Chapter 4

Univariate Integrals

4.1 Introduction

In this chapter we derive two important results about the asymptotic expansion of univariate integrals. We consider only Laplace integrals

$$\int_0^\infty e^{-\lambda t} h(t)\, dt \tag{4.1}$$

and Laplace type integrals

$$\int_D e^{\lambda f(t)} h(t)\, dt. \tag{4.2}$$

Here D is a subset of $\mathbb{R}$.

4.2 Watson's Lemma

In this section an asymptotic approximation for Laplace transforms is derived. This result was proved by Watson in [138].

Given is the Laplace transform

$$I(\lambda) = \int_0^\infty e^{-\lambda t} f(t)\, dt \tag{4.3}$$

of a function $f(t)$. Assume that for this function

$$f(t) = O(e^{at}),\ t \to \infty, \tag{4.4}$$

$$f(t) \sim \sum_{m=0}^{\infty} c_m t^{a_m},\ t \to 0+. \tag{4.5}$$

with $-1 < a_0 < a_1 < a_2 < \ldots < a_n \to \infty$. The a_m's are not necessarily integers.

Lemma 35 (Watson's Lemma) *Given is a locally integrable function $f(t)$ on $(0,\infty)$ bounded for finite t and (4.4) and (4.5) hold. Then, as $\lambda \to \infty$*

$$I(\lambda) \sim \sum_{m=0}^{\infty} c_m \int_0^{\infty} e^{-\lambda t} t^{a_m}\, dt = \sum_{m=0}^{\infty} \frac{c_m \Gamma(a_m+1)}{\lambda^{a_m+1}}. \tag{4.6}$$

PROOF: Let R be a fixed positive number. Then we write

$$I(\lambda) = \underbrace{\int_0^R e^{-\lambda t} f(t)\, dt}_{=I_1(\lambda)} + \underbrace{\int_R^{\infty} e^{-\lambda t} f(t)\, dt}_{=I_2(\lambda)}. \tag{4.7}$$

From equation (4.4) follows that there is a constant K such that $|f(t)| \le Ke^{at}$ for all $t > R$. Hence we get for $\lambda > a$ always

$$|I_2(\lambda)| \le \int_R^{\infty} e^{-\lambda t} |f(t)|\, dt \tag{4.8}$$

$$\begin{aligned} \le \int_R^{\infty} e^{-\lambda t} K e^{at}\, dt &= K \int_R^{\infty} e^{-(\lambda-a)t} dt = \frac{K}{\lambda - a} e^{-(\lambda-a)R} \\ &= \frac{e^{-\lambda R}}{\lambda} \left[\frac{\lambda K}{\lambda - a} e^{-aR} \right] = o(e^{-\lambda R}). \end{aligned}$$

The last relation follows, since the term in the square brackets remains bounded as $\lambda \to \infty$.

For each natural number N we can write using equation (4.5)

$$f(t) = \sum_{m=0}^{N} c_m t^{a_m} + \rho_N(t) \tag{4.9}$$

with $\rho_n(t) = O(t^{a_{N+1}})$ as $t \to 0+$. Then we write the integral $I_1(\lambda)$ in the form

$$I_1(\lambda) = \sum_{m=0}^{N} c_m \int_0^R t^{a_m} e^{-\lambda t}\, dt + \int_0^R \rho_N(t) e^{-\lambda t}\, dt. \tag{4.10}$$

For these integrals we get then

$$\begin{aligned} \int_0^R t^{a_m} e^{-\lambda t} dt &= \int_0^{\infty} t^{a_m} e^{-\lambda t}\, dt - \int_R^{\infty} t^{a_m} e^{-\lambda t}\, dt \\ &= \frac{\Gamma(1+a_m)}{\lambda^{1+a_m}} + O(e^{-\lambda R}),\ \lambda \to \infty. \end{aligned} \tag{4.11}$$

Now we have that

$$I(\lambda) = \sum_{m=0}^{N} c_m \Gamma(a_m + 1)\lambda^{-(1+a_m)} + \int_0^R \rho_N(t) e^{-\lambda t}\, dt + O(e^{-\lambda R}). \tag{4.12}$$

From (4.5) and since $\rho_N(t)$ is bounded on $(0, R)$, there is a constant K_N such that

$$|\rho_N(t)| \le K_N t^{a_{N+1}} \tag{4.13}$$

for all $t \in (0, R]$. Then we obtain an upper bound

$$\left| \int_0^R \rho_N(t) e^{-\lambda t}\, dt \right| \le K_N \int_0^\infty t^{a_{N+1}} e^{-\lambda t}\, dt = K_N \frac{\Gamma(a_{N+1}+1)}{\lambda^{a_{N+1}+1}} . \tag{4.14}$$

With the equations (4.8), (4.11) and (4.14) the final result is obtained. □

4.3 The Laplace Method for Univariate Functions

For Laplace type functions it is possible to derive asymptotic approximations by studying the structure of the functions in the integral near the global maximum points of the function in the exponent.

Given is a function $I(\lambda)$ of the real variable λ in the form of an integral

$$I(\lambda) = \int_\alpha^\beta h(x) \exp(\lambda f(x))\, dx. \tag{4.15}$$

Here $\alpha \in I\!R$ and $[\alpha, \beta)$ is a finite or semi-infinite interval, $h(x)$ and $f(x)$ are real functions and λ is a real parameter. The Laplace method gives asymptotic approximations for $I(\lambda)$ as $\lambda \to \infty$.

To derive this approximation, one studies the functions $h(x)$ and $f(x)$ in the neighborhood of the points where the function f has its global maximum $\max_{\alpha \le x \le \beta} f(x)$. As $\lambda \to \infty$ the integrand is concentrated more and more around these points.

Without loss of generality it can be assumed that the maximum occurs at one point x_0 only. If there are several points, the interval $[\alpha, \beta)$ is split up into several subintervals such that in each of them is exactly one point at which the global maximum occurs.

The following result is from [57], p. 37.

Theorem 36 *Assume that the following conditions are fulfilled:*

1. *$f(x)$ is a real function on the finite or semi-infinite interval $[\alpha, \beta)$ and in an interval $(\alpha, \alpha+\epsilon]$ with $\epsilon > 0$ it is continuously differentiable and*

$$\sup_{\alpha+\epsilon \le x \le \beta} f(x) \le f(\alpha) - \delta, \text{ with } \delta > 0. \tag{4.16}$$

For the derivative $f'(x)$ we have that

$$\begin{aligned} f'(x) &< 0 \text{ for all } x \in (\alpha, \alpha+\epsilon], & (4.17)\\ f'(x) &= -a(x-\alpha)^{r-1} + o((x-\alpha)^{r-1}) \text{ with } r > 0. & (4.18) \end{aligned}$$

2. *$h(x)$ is a real and continuous function on the interval $[\alpha, \beta)$ with*

$$h(x) = b(x-\alpha)^{s-1} + o((x-\alpha)^{s-1}) \text{ with } s > 0. \tag{4.19}$$

3. *The integral*

$$\int_\alpha^\beta |h(x)| \exp(f(x))\ dx \tag{4.20}$$

is finite.

Then the integrals

$$I(\lambda) = \int_\alpha^\beta h(x) \exp(\lambda f(x))\ dx \tag{4.21}$$

with $\lambda \ge 1$ are all finite and have the asymptotic approximation

$$I(\lambda) \sim \frac{b}{r} \Gamma\left(\frac{s}{r}\right) \left(\frac{r}{a}\right)^{\frac{s}{r}} \lambda^{-\frac{s}{r}} \exp(\lambda f(\alpha)),\ \lambda \to \infty. \tag{4.22}$$

PROOF: For simplicity we assume that $f(\alpha) = 0$. To show that all integrals $I(\lambda)$ are finite, we get an upper bound for them

$$|I(\lambda)| \le \int_\alpha^\beta |h(x)| \exp(\lambda f(x))\ dx \tag{4.23}$$

$$\le \underbrace{\int_\alpha^{\alpha+\epsilon} |h(x)| \exp(\lambda f(x))\ dx}_{=I_1(\lambda)} + \underbrace{\int_{\alpha+\epsilon}^\beta |h(x)| \exp(\lambda f(x))\ dx}_{=I_2(\lambda)}.$$

The first integral is finite. For the second we get using equation (4.16) that

$$|I_2(\lambda)| \leq \int_{\alpha+\epsilon}^{\beta} |h(x)| \exp\left(f(x) + (\lambda - 1)(f(\alpha) - \delta)\right) \, dx \tag{4.24}$$

$$= \exp(-\delta(\lambda - 1)) \int_{\alpha+\epsilon}^{\beta} |h(x)| \exp\left(f(x)\right) \, dx = O(\exp(-\lambda\delta)).$$

But since due to assumption 3) the last integral is finite, we find that $|I(\lambda)| < \infty$ for all $\lambda > 0$.

In the first integral we make a substitution $x \to u = -f(x)$. We have then

$$\frac{d\,u}{d\,x} = -f'(x), \tag{4.25}$$

and

$$\frac{d\,x}{d\,u} = -\frac{1}{f'(x)}. \tag{4.26}$$

Now we can write the first integral in the form

$$I_1(\lambda) = \int_0^{f(\alpha+\epsilon)} \underbrace{\left[h(x(u))\frac{d\,x}{d\,u}\right]}_{=k(u)} \exp(\lambda u) \, du. \tag{4.27}$$

We have

$$u = -f(x) = -\int_{\alpha}^{x} f'(y) \, dy. \tag{4.28}$$

Using the asymptotic form of f' we get by integrating that

$$u \sim \frac{a}{r}(x - \alpha)^r, \; x \to \alpha. \tag{4.29}$$

Therefore

$$x - \alpha \sim \left(\frac{ur}{a}\right)^{1/r}, \; u \to 0. \tag{4.30}$$

For the function in the square brackets in equation (4.27) we find then the expansion as $u \to 0$

$$h(x(u))\frac{d\,x}{d\,u} \sim \frac{b}{a}(x - \alpha)^{s-r}. \tag{4.31}$$

and so we get

$$k(u) \sim \frac{b}{a}\left(\frac{ur}{a}\right)^{s/r-1}. \tag{4.32}$$

Using now Watson's lemma gives the result. □

From this result some special cases can be derived.

Corollary 37 *Let f and h be continuous functions on a finite interval $[\alpha, \beta]$. For some important special cases the last theorem gives :*
a) If the global maximum of $f(x)$ occurs only at the point $x_0 \in (\alpha, \beta)$, $h(x_0) \neq 0$, $f(x)$ is near x_0 twice continuously differentiable and $f''(x_0) < 0$, then

$$I(\lambda) \sim h(x_0) \exp(\lambda f(x_0)) \sqrt{\frac{2\pi}{\lambda |f''(x_0)|}}, \quad \lambda \to \infty. \tag{4.33}$$

b) If the global maximum occurs only at α, $h(\alpha) \neq 0$ and $f(x)$ is near α continuously differentiable with $f'(\alpha) < 0$, then

$$I(\lambda) \sim h(\alpha) \exp(\lambda f(\alpha)) \frac{1}{\lambda |f'(\alpha)|}, \quad \lambda \to \infty. \tag{4.34}$$

c) If the global maximum occurs only at α, $h(\alpha) \neq 0$ and $f(x)$ is twice continuously differentiable near α with $f'(\alpha) = 0$ and $f''(\alpha) < 0$, then

$$I(\lambda) \sim h(\alpha) \exp(\lambda f(\alpha)) \sqrt{\frac{\pi}{2\lambda |f''(\alpha)|}}, \quad \lambda \to \infty. \tag{4.35}$$

PROOF: The results follow directly from the last theorem. □

For all these integrals it is possible to derive asymptotic expansions in the sense of Poincaré if higher derivatives of f and h exist at the global maximum points (see [15], chapter 5 and [140], p. 58).

Chapter 5

Multivariate Laplace Type Integrals

5.1 Introduction

An important part of asymptotic analysis is the derivation of approximation methods for integrals depending on one or more parameters. In the last chapter some results about univariate integrals were given, here we will consider multivariate integrals. The two most important types of integrals are

1. Laplace type integrals

$$I(\lambda) = \int_F h(\boldsymbol{x}) e^{\lambda f(\boldsymbol{x})}\, d\boldsymbol{x}$$

2. and Fourier type integrals

$$J(\lambda) = \int_F h(\boldsymbol{x}) e^{i\lambda f(\boldsymbol{x})}\, d\boldsymbol{x}.$$

Here F is a subset of $\mathbb{R}^n$, $h(\boldsymbol{x})$ and $f(\boldsymbol{x})$ are functions defined on F and λ is a real parameter. Here the asymptotic behavior of the functions $I(\lambda)$ or $J(\lambda)$ is studied as λ approaches infinity or zero. In the following we will consider only Laplace type integrals. For these integrals the asymptotic behavior is dominated by the structure of the functions near the global maximum points of f with respect to F. In the following we will consider only Laplace type integrals.

In the case of Fourier integrals also other stationary points of this functions may be important for the asymptotic behavior of the integrals, therefore the structure of the functions near all points with $\nabla f(\boldsymbol{x}) = \boldsymbol{o}$ has to be considered. Such results can be found in [15], chap. 8.4.

5.2 Basic Results

In this section we consider asymptotic approximations for multivariate Laplace type integrals as $\beta \to \infty$

$$I(\beta) = \int_F h(\boldsymbol{x}) \exp(\beta^2 f(\boldsymbol{x}))\, d\boldsymbol{x}. \tag{5.1}$$

with $F \subset I\!R^n$, $h(\boldsymbol{x})$ a continuous function and $f(\boldsymbol{x})$ a twice continuously differentiable function. Here β is a real parameter. We set here as parameter β^2 instead of λ deviating from the usual terminology, since in the applications in the following chapters for normal integrals the parameter appears in this form.

In the years 1948-1966 the Laplace method was generalized to approximations for multivariate integrals. The first result is given in the article of Hsu [76] in 1948. Here the most simple case as described in theorem 41, p. 56 is treated. The most extensive monography about the Laplace method for bivariate integrals is from J. Focke [63], published 1953. Fulks/Sather [66] proved a theorem which allows that the function h is zero at the maximum point of f or has a slight singularity at this point, see theorem 43, p. 60. Jones [79] then in the year 1966 proved the theorem 46, p. 67 about boundary maxima. In the following we will consider only finite-dimensional spaces; some results about Laplace integrals in infinite-dimensional spaces are given in [53], [55] and [54]. In these papers a similar problem, i.e. the asymptotic form of functionals of probability measures, which converge towards Gaussian probability measures, is studied.

The asymptotic behavior of $I(\beta)$ as $\beta \to \infty$ can be studied with similar methods as in the univariate case. But here some additional problems appear and due to this there is no complete theory for the asymptotics of these integrals until now. In this report a summary of results will be given, which should be sufficient for most problems in applications.

In principle the asymptotic behavior of these integrals is determined by the structure of the functions f and h and of the integration domain F in the neighborhood of the points or sets where the function f achieves its global maximum with respect to F.

In Sirovich [127] in chapter 2.8 the different types of maximum points are described:

1. points in the interior of F (Type I),
2. points at the boundary of F with a smooth boundary in the neighborhood (Type II),
3. points at the boundary of F where the boundary is non-smooth (Type III).

Depending on the type of the maximum point different asymptotic approximations are obtained. In general these approximations depend on the values of the functions f, g and h and their first two derivatives at the maximum points. Only if these vanish, higher derivatives must be computed (see [140], chap. VIII, 5). Then the first non-vanishing derivatives determine the form of the asymptotics.

The standard results for integrals with a maximum in the interior, i.e. theorem 41, p. 56 and for a maximum at the boundary, theorem 46, p. 67, can be found in the textbooks [15], chap. 8, [140], chap. IX and [127], chap. 2.8.

Additional to the standard results in this report we also treat the case that the maximum of the function is a submanifold of the boundary of the integration domain and that there is an additional function depending on the parameter β.

To avoid unnecessary technical details, we first prove a lemma that shows that under slight regularity conditions it is often possible to restrict the integration domain to a compact set.

The following lemma gives conditions under which it is possible to replace an integral over a non-compact domain by an integral over a compact domain without changing its asymptotic behavior. This is a generalization of a result of M. Hohenbichler, published in [33].

Lemma 38 (Compactification Lemma) *Given is a closed set $F \subset \mathbb{R}^n$ and two continuous functions f and $h : \mathbb{R}^n \to \mathbb{R}$. Assume further that:*

1. *The set $M = \{\boldsymbol{y} \in F; f(\boldsymbol{y}) = \max_{\boldsymbol{x} \in F} f(\boldsymbol{x})\}$ is compact,*

2. $\int_F |h(\boldsymbol{y})| e^{f(\boldsymbol{y})} d\boldsymbol{y} < \infty$,

3. *For every neighborhood V of M : $\sup\{f(\boldsymbol{y}); \boldsymbol{y} \in F \setminus V\} < \max_{\boldsymbol{x} \in F} f(\boldsymbol{x})$,*

4. *There exists a neighborhood U of M such that for all $\boldsymbol{x} \in U$ always $h(\boldsymbol{x}) > 0$ (or $h(\boldsymbol{x}) < 0$).*

5. *For all neighborhoods V of M always*

$$\int_{F \cap V} d\boldsymbol{x} > 0. \tag{5.2}$$

Then for all $\lambda > 1$

$$\int_F |h(\boldsymbol{x})| e^{\lambda f(\boldsymbol{x})} \, d\boldsymbol{x} < \infty, \tag{5.3}$$

and for all neighborhoods V of M holds that

$$\int_F h(\boldsymbol{x}) e^{\lambda f(\boldsymbol{x})} \, d\boldsymbol{x} \sim \int_{F \cap V} h(\boldsymbol{x}) e^{\lambda f(\boldsymbol{x})} \, d\boldsymbol{x}, \ \lambda \to \infty. \tag{5.4}$$

PROOF: Let be $m = \sup_{x \in F} f(x)$ and δ, ϵ be positive constants. For $\lambda \geq 1$ and $f(y) \geq m - \delta$ we have then

$$e^{(\lambda-1)f(y)} \geq e^{(\lambda-1)(m-\delta)} \tag{5.5}$$

or

$$e^{\lambda f(y)} \geq e^{(\lambda-1)(m-\delta)} e^{f(y)}. \tag{5.6}$$

In the same way for $\lambda \geq 1$ and $f(y) \leq m - \epsilon$ we find

$$e^{\lambda f(y)} \leq e^{(\lambda-1)(m-\epsilon)} e^{f(y)}. \tag{5.7}$$

Since $f(y) \leq m$ for all $y \in F$ we get from equation (5.7) in the limit as $\epsilon \to 0$ that

$$\int_F |h(y)| e^{\lambda f(y)} \, dy \leq e^{(\lambda-1)m} \int_F |h(y)| e^{f(y)} \, dy < \infty.$$

This proves the first part of the lemma.

Now we assume that $h(y) > 0$ for all $y \in V$ with V being a neighborhood of M. In the following we use the notations

$$G(\lambda, V) = \int_{F \cap V} h(y) e^{\lambda f(y)} \, dy \quad , \quad G_a(\lambda, V) = \int_{F \cap V} |h(y)| e^{\lambda f(y)} \, dy, \tag{5.8}$$

$$\bar{G}(\lambda, V) = \int_{F \setminus V} h(y) e^{\lambda f(y)} \, dy \quad , \quad \bar{G}_a(\lambda, V) = \int_{F \setminus V} |h(y)| e^{\lambda f(y)} \, dy. \tag{5.9}$$

First we prove the proposition (+):

(+) If V is a neighborhood of M with $h(y) > 0$ for all $y \in V$, then for all neighborhoods W of M with $V \subset W$ always

$$G(\lambda, V) \sim G(\lambda, W), \ \lambda \to \infty. \tag{5.10}$$

Let $\epsilon = m - \sup\{f(y); y \in F \setminus V\}$. Since f is continuous, there exists a neighborhood $V_1 \subset V$ of M such that for $y \in V_1$ always

$$f(y) \geq m - \epsilon/2. \tag{5.11}$$

Since we assumed that $h(y) > 0$ for $y \in V$, we get with equation (5.5) that

$$|G(\lambda, V_1)| = G_a(\lambda, V_1) \geq e^{(\lambda-1)(m-\epsilon/2)} G_a(1, V_1). \tag{5.12}$$

Here is $G_a(1, V_1) > 0$ due to assumption 5.

From equation (5.7) follows further that

$$\bar{G}_a(\lambda, V) \leq e^{(\lambda-1)(m-\epsilon)} \bar{G}_a(1, V). \tag{5.13}$$

with $\bar{G}_a(1, V) < \infty$ due to assumption 2.

From the relations (5.12) and (5.13) follows

$$\frac{\bar{G}_a(\lambda, V)}{G(\lambda, V_1)} \leq \frac{e^{(\lambda-1)(m-\epsilon)}\bar{G}_a(1, V)}{e^{(\lambda-1)(m-\epsilon/2)}G_a(1, V)} \tag{5.14}$$
$$= e^{-\lambda\epsilon/2}\frac{\bar{G}_a(1, V)}{G_a(1, V)} \to 0, \ \lambda \to \infty.$$

But since $h(y) > 0$ for all $y \in V$, we have that

$$0 \leq G(\lambda, V_1) \leq G(\lambda, V) \tag{5.15}$$

and further also for an arbitrary neighborhood W with $V \subset W$ that

$$0 \leq G_a(\lambda, W \setminus V) \leq G_a(\lambda, \mathbb{R}^n \setminus V) = \bar{G}_a(\lambda, V). \tag{5.16}$$

From the equations (5.14), (5.15) und (5.16) we get finally that

$$\lim_{\lambda \to \infty} \frac{G_a(\lambda, W \setminus V)}{G(\lambda, V)} = 0. \tag{5.17}$$

This gives statement (+).

Let now $\tilde{V}$ be an arbitrary neighborhood of M. Then there exists a neighborhood $V_1 \subset \tilde{V}$ of M with $h(y) > 0$ for all $y \in V_1$ due to assumption 4.

From (+) we get then

$$G(\lambda, V_1) \sim G(\lambda, V), \ \lambda \to \infty \text{ and } G(\lambda, V_1) \sim G(\lambda, \mathbb{R}^n), \ \lambda \to \infty.$$

From these two asymptotic equations we obtain the second part of the lemma.
□

5.3 Interior Maxima

We consider now the case that the maximum of the function in the argument of the exponent of a Laplace type integral occurs at one point in the interior of the integration domain. The first theorem is a well known standard result.

First two lemmas are proved which are needed in the following to get estimates of the integrals.

Lemma 39 *Let β be a positive real parameter. Then for all constants $\epsilon, k > 0$ always*

$$\int_{|x|>\beta\epsilon} |x| \exp(-k|x|^2)\, dx \to 0, \ \beta \to \infty. \tag{5.18}$$

PROOF: By using spherical coordinates we get that

$$\int_{|x|>\beta\epsilon} |x| \exp(-k|x|^2)\, dx = \frac{2\pi^{n/2}}{\Gamma(n/2)} \int_{\rho>\beta\epsilon} \rho^n \exp(-k\rho^2)\, d\rho \to 0 \tag{5.19}$$

as $\beta \to \infty$. □

Lemma 40 *Let $F \subset \mathbb{R}^n$ be a compact set with the origin in its interior. Further is given a twice continuously differentiable function $f : F \to \mathbb{R}$. Assume that further:*

1. *$f(\mathbf{o}) > f(\boldsymbol{x})$ for all $\boldsymbol{x} \in F$ with $\boldsymbol{x} \neq \mathbf{o}$,*
2. *The Hessian $\boldsymbol{H}_f(\mathbf{o})$ is negative definite.*

Then there exists a constant $k > 0$ such that for all $\boldsymbol{x} \in F$ always

$$f(\boldsymbol{x}) \leq f(\mathbf{o}) - k|\boldsymbol{x}|^2. \tag{5.20}$$

PROOF: We assume that $f(\mathbf{o}) = 0$. If we now assume that for all $k > 0$ there is a point $\boldsymbol{x} \in F$ with $-k < f(\boldsymbol{x})|\boldsymbol{x}|^{-2} < 0$, then there exists a sequence $(\boldsymbol{x}^{(m)})$ of points in F with $-m^{-1} \leq f(\boldsymbol{x}^{(m)})|\boldsymbol{x}^{(m)}|^{-2} < 0$. Since F is compact, this sequence has a convergent subsequence which converges towards a point $\boldsymbol{x}_0 \in F$. Since f is continuous, we have then that $f(\boldsymbol{x}_0) = 0$. From assumption 1. follows then that $\boldsymbol{x}_0 = \mathbf{o}$. Due to assumption 2. the Hessian $\boldsymbol{H}_f(\mathbf{o})$ is negative definite and making a Taylor expansion around $\mathbf{o}$ we get with $\nabla f(\mathbf{o}) = \mathbf{o}$ that

$$f(\boldsymbol{y}) = f(\mathbf{o}) + (\nabla f(\mathbf{o}))^T \boldsymbol{y} + \frac{1}{2}\boldsymbol{y}^T \boldsymbol{H}_f(\vartheta \boldsymbol{y})\boldsymbol{y} = \frac{1}{2}\boldsymbol{y}^T \boldsymbol{H}_f(\vartheta \boldsymbol{y})\boldsymbol{y} \tag{5.21}$$

with $0 \leq \vartheta \leq 1$.

Since f is twice continuously differentiable, $\boldsymbol{H}_f(\vartheta \boldsymbol{y}^{(m)}) \to \boldsymbol{H}_f(\mathbf{o})$ if $\boldsymbol{y}^{(m)} \to \mathbf{o}$. Therefore for all sequences $(\boldsymbol{y}^{(m)})$ with $\boldsymbol{y}^{(m)} \to \mathbf{o}$ always

$$\limsup_{m \to \infty} f(\boldsymbol{y}^{(m)})|\boldsymbol{y}^{(m)}|^{-2} \leq \frac{1}{2}\lambda_0 < 0, \tag{5.22}$$

with $\lambda_0 < 0$ being the largest eigenvalue of the Hessian $\boldsymbol{H}_f(\mathbf{o})$ and so such a sequence $(\boldsymbol{x}^{(m)})$ can not exist. Therefore such a constant $k > 0$ exists. □

Theorem 41 *Given is a compact set $F \subset \mathbb{R}^n$, a twice continuously differentiable function $f : \mathbb{R}^n \to \mathbb{R}$ and a continuous function $h : \mathbb{R}^n \to \mathbb{R}$. Let be defined*

$$I(\beta) = \int_F h(\boldsymbol{x}) \exp(\beta^2 f(\boldsymbol{x}))\, d\boldsymbol{x}. \tag{5.23}$$

If the function f achieves its global maximum with respect to F only at the point $\boldsymbol{x}^*$ in the interior of F and if the Hessian $\boldsymbol{H}_f(\boldsymbol{x}^*)$ is negative definite, the following asymptotic equation is valid

$$I(\beta) \sim (2\pi)^{n/2} \frac{h(\boldsymbol{x}^*)}{\sqrt{|\det(\boldsymbol{H}_f(\boldsymbol{x}^*))|}} \cdot e^{\beta^2 f(\boldsymbol{x}^*)} \beta^{-n}, \quad \beta \to \infty. \tag{5.24}$$

PROOF: A proof is given in [15], p. 332-4, [59], p. 74-75 and [140], p. 495.

The idea of the proof is to approximate first the functions f and h by Taylor expansions at the maximum point and then to show that the integral over these approximating Taylor expansions is asymptotically equivalent to the integral over the functions. Near $\boldsymbol{x}^*$ the function f can be approximated by its second order Taylor expansion at $\boldsymbol{x}^*$.

Since f is twice differentiable and at $\boldsymbol{x}^*$ there is an extremum, the first derivatives vanish, i.e. $\nabla f(\boldsymbol{x}^*) = \mathbf{o}$ and so we get that

$$f(\boldsymbol{x}) \sim f(\boldsymbol{x}^*) + \frac{1}{2}(\boldsymbol{x} - \boldsymbol{x}^*)^T \boldsymbol{H}_f(\boldsymbol{x}^*)(\boldsymbol{x} - \boldsymbol{x}^*). \tag{5.25}$$

For h it is sufficient to replace the function $h(\boldsymbol{x})$ by its value at $h(\boldsymbol{x}^*)$. Hence we have near $\boldsymbol{x}^*$ the approximation

$$h(\boldsymbol{x})e^{\beta^2 f(\boldsymbol{x})} \sim h(\boldsymbol{x}^*)\exp\left[\beta^2(f(\boldsymbol{x}^*) + \frac{1}{2}(\boldsymbol{x} - \boldsymbol{x}^*)^T \boldsymbol{H}_f(\boldsymbol{x}^*)(\boldsymbol{x} - \boldsymbol{x}^*))\right]. \tag{5.26}$$

Let now $U \subset \mathbb{R}^n$ be a neighborhood of $\boldsymbol{x}^*$. One can show by estimating the differences that

$$\int_F h(\boldsymbol{x})e^{\beta^2 f(\boldsymbol{x})}\, d\boldsymbol{x} \tag{5.27}$$

$$\sim \int_U h(\boldsymbol{x}^*)\exp\left[\beta^2(f(\boldsymbol{x}^*) + \frac{1}{2}(\boldsymbol{x} - \boldsymbol{x}^*)^T \boldsymbol{H}_f(\boldsymbol{x}^*)(\boldsymbol{x} - \boldsymbol{x}^*))\right]\, d\boldsymbol{x}$$

$$\sim \int_{\mathbb{R}^n} h(\boldsymbol{x}^*)\exp\left[\beta^2(f(\boldsymbol{x}^*) + \frac{1}{2}(\boldsymbol{x} - \boldsymbol{x}^*)^T \boldsymbol{H}_f(\boldsymbol{x}^*)(\boldsymbol{x} - \boldsymbol{x}^*))\right]\, d\boldsymbol{x}$$

$$= h(\boldsymbol{x}^*)e^{\beta^2 f(\boldsymbol{x}^*)} \int_{\mathbb{R}^n} \exp\left[\frac{\beta^2}{2}(\boldsymbol{x} - \boldsymbol{x}^*)^T \boldsymbol{H}_f(\boldsymbol{x}^*)(\boldsymbol{x} - \boldsymbol{x}^*)\right]\, d\boldsymbol{x}$$

But for the last integral we have (see lemma 26, p. 30) that

$$= (2\pi)^{n/2} \frac{h(\boldsymbol{x}^*)}{\sqrt{|\det(\boldsymbol{H}_f(\boldsymbol{x}^*))|}} \cdot e^{\beta^2 f(\boldsymbol{x}^*)}\beta^{-n}. \tag{5.28}$$

This gives the asymptotic relation in the theorem.

We give now an exact proof. We assume that $\boldsymbol{x}^* = \mathbf{o}$ and $f(\mathbf{o}) = 0$. First we define the functions

$$h_\beta(\boldsymbol{x}) = h(\beta^{-1}\boldsymbol{x})\exp(\beta^2 f(\beta^{-1}\boldsymbol{x})). \tag{5.29}$$

By defining the functions h and f for all $\boldsymbol{x} \in \mathbb{R}^n \setminus F$ by $f(\boldsymbol{x}) = h(\boldsymbol{x}) = 0$, the functions h_β are defined on $\mathbb{R}^n$. Then we have

$$\beta^n I(\beta) = \beta^n \int_{\mathbb{R}^n} h(\boldsymbol{y})\exp(\beta^2 f(\boldsymbol{y}))\, d\boldsymbol{y}. \tag{5.30}$$

Making the substitution $\boldsymbol{y} \to \boldsymbol{x} = \beta \boldsymbol{y}$ gives

$$\beta^n I(\beta) = \int_{\mathbb{R}^n} h(\beta^{-1}\boldsymbol{x}) \exp(\beta^2 f(\beta^{-1}\boldsymbol{x}))\, d\boldsymbol{x} = \int_{\mathbb{R}^n} h_\beta(\boldsymbol{x})\, d\boldsymbol{x}. \tag{5.31}$$

We show now that we can apply the Lebesgue theorem for the limit of the integrals over the functions $h_\beta(\boldsymbol{x})$. First we note that lemma 40, p. 55 ensures that there is a constant $k > 0$ such that

$$|h_\beta(\boldsymbol{x})| \le \max_{\boldsymbol{x}\in D} |h(\boldsymbol{x})| \exp(-\beta^2(k\beta^{-2}|\boldsymbol{x}|^2)) = \max_{\boldsymbol{x}\in D} |h(\boldsymbol{x})| \exp(-k|\boldsymbol{x}|^2). \tag{5.32}$$

Since the last function is integrable we have an integrable upper bound for all these functions and we can apply the Lebesgue theorem if we find the limit function. We have then, since h is continuous

$$\lim_{\beta\to\infty} h(\beta^{-1}\boldsymbol{x}) = h(\mathbf{o}). \tag{5.33}$$

Since $\mathbf{o}$ is in the interior of F, for all $\boldsymbol{x}$ there is a β_0 such that for all $\beta > \beta_0$ the points $\beta^{-1}\boldsymbol{x}$ lie in F and therefore then the function $f(\beta^{-1}\boldsymbol{x})$ is twice differentiable with respect to β for these values. For the function $\beta^2 f(\beta^{-1}\boldsymbol{x})$ we get using l'Hospital's rule twice

$$\begin{aligned}
\lim_{\beta\to\infty} \beta^2 f(\beta^{-1}\boldsymbol{x}) &= \lim_{\beta\to\infty} \frac{f(\beta^{-1}\boldsymbol{x})}{\beta^{-2}} \qquad (5.34)\\
= \lim_{\beta\to\infty} \frac{-\beta^{-2}\sum_{i=1}^n f^i(\beta^{-1}\boldsymbol{x})x_i}{-2\beta^{-3}} &= \lim_{\beta\to\infty} \frac{-\sum_{i=1}^n f^i(\beta^{-1}\boldsymbol{x})x_i}{-2\beta^{-1}}\\
= \lim_{\beta\to\infty} \frac{-\beta^{-2}\sum_{i,j=1}^n f^{ij}(\beta^{-1}\boldsymbol{x})x_i x_j}{-2\beta^{-2}} &= \lim_{\beta\to\infty} \frac{1}{2}\sum_{i,j=1}^n f^{ij}(\beta^{-1}\boldsymbol{x})x_i x_j\\
= \frac{1}{2}\sum_{i,j=1}^n f^{ij}(\mathbf{o})x_i x_j &= \frac{1}{2}\boldsymbol{x}^T \boldsymbol{H}_f(\mathbf{o})\boldsymbol{x}.
\end{aligned}$$

So we find for the limit

$$\lim_{\beta\to\infty} \beta^n I(\beta) = \int_{\mathbb{R}^n} h(\mathbf{o}) \exp\left(\frac{1}{2}\boldsymbol{x}^T \boldsymbol{H}_f(\mathbf{o})\boldsymbol{x}\right) d\boldsymbol{x}. \tag{5.35}$$

Using lemma 26, p. 30 we get finally the result

$$\lim_{\beta\to\infty} \beta^n I(\beta) = (2\pi)^{n/2} h(\mathbf{o})|\boldsymbol{H}_f(\mathbf{o})|^{-1/2}. \tag{5.36}$$

□

The meaning of the approximation above is that asymptotically as $\beta \to \infty$ the function $\exp\left[\beta^2(f(\boldsymbol{x}) - f(\boldsymbol{x}^*))\right]$ near $\boldsymbol{x}^*$ is proportional to the density of a normal distribution $\mathrm{N}(\mathbf{o}, -\beta^{-2}\boldsymbol{H}_f^{-1}(\boldsymbol{x}^*))$. Showing that the integration domain can be replaced by $\mathbb{R}^n$ without changing the essential asymptotic behavior, an integral over a normal density is obtained.

In the following theorem the function $h(\boldsymbol{x})$ is vanishing at the maximum point.

Theorem 42 *Given is a multivariate integral $I(\beta)$ depending on a real parameter β*

$$I(\beta) = \int_F g_0(\boldsymbol{x})|g_1(\boldsymbol{x})| \exp(\beta^2 f(\boldsymbol{x}))\, d\boldsymbol{x}. \tag{5.37}$$

Further the following conditions hold:

1. *F is a compact set in $\mathbb{R}^n$,*
2. *$f : \mathbb{R}^n \to \mathbb{R}$ is a twice continuously differentiable function with*
   (a) *$f(\boldsymbol{x}) < f(\boldsymbol{x}^*)$ for all $\boldsymbol{x} \in D$ with $\boldsymbol{x} \neq \boldsymbol{x}^*$,*
   (b) *The Hessian $\boldsymbol{H}_f(\boldsymbol{x}^*)$ is negative definite,*
   (c) *$g_0 : \mathbb{R}^n \to \mathbb{R}$ is a continuous function with $g_0(\boldsymbol{x}^*) \neq 0$,*
   (d) *$g_1 : \mathbb{R}^n \to \mathbb{R}$ is a continuously differentiable function vanishing at $\boldsymbol{x}^*$, i.e. $g_1(\boldsymbol{x}^*) = 0$.*

Then the integral $I(\beta)$ has the following asymptotic approximation as $\beta \to \infty$

$$I(\beta) \sim 2(2\pi)^{(n-1)/2} g_0(\boldsymbol{x}^*) \left| \frac{(\nabla g_1(\boldsymbol{x}^*))^T \boldsymbol{H}_f^{-1}(\boldsymbol{x}^*) \nabla g_1(\boldsymbol{x}^*)}{\det(\boldsymbol{H}_f(\boldsymbol{x}^*))} \right|^{1/2} \cdot \frac{e^{\beta^2 f(\boldsymbol{x}^*)}}{\beta^{n+1}}. \tag{5.38}$$

PROOF: We assume that $\boldsymbol{x}^* = \mathbf{o}$, $f(\mathbf{o}) = 0$ and F is convex. We use the Lebesgue convergence theorem 8, p. 12. We define the function $h_\beta(\boldsymbol{x})$ by

$$h_\beta(\boldsymbol{x}) = g_0(\beta^{-1}\boldsymbol{x})(\beta|g_1(\beta^{-1}\boldsymbol{x})|) \exp(\beta^2 f(\beta^{-1}\boldsymbol{x})). \tag{5.39}$$

for $\boldsymbol{x} \in F$ and zero elsewhere.

Since F is convex, we have that $|\beta g_1(\beta^{-1}\boldsymbol{x})| \leq K_1|\boldsymbol{x}|$ with $K_1 = \max_{\boldsymbol{x}\in F} |\nabla g_1(\boldsymbol{x})|$. Therefore we can find again an upper bound for all functions $|h_\beta(\boldsymbol{x})|$.

Then we have

$$I(\beta) = \beta^{-(n+1)} \int_{\mathbb{R}^n} h_\beta(\boldsymbol{x})\, d\boldsymbol{x}. \tag{5.40}$$

For the function $h_\beta(\boldsymbol{x})$ we have that

$$\lim_{\beta\to\infty} g_0(\beta^{-1}\boldsymbol{x}) = g_0(\mathbf{o}). \tag{5.41}$$

Applying l'Hospital's rule, we get

$$\begin{aligned} \lim_{\beta\to\infty} \beta|g_1(\beta^{-1}\boldsymbol{x})| &= \lim_{\beta\to\infty} \frac{\frac{dg_1(\beta^{-1}\boldsymbol{x})}{d\beta}}{\frac{d\beta^{-1}}{d\beta}} = \frac{-\beta^{-2}(\nabla g_1(\beta^{-1}\boldsymbol{x}))^T\boldsymbol{x}}{-\beta^{-2}} \\ &= \lim_{\beta\to\infty} (\nabla g_1(\beta^{-1}\boldsymbol{x}))^T\boldsymbol{x} = (\nabla g_1(\mathbf{o}))^T\boldsymbol{x}. \end{aligned} \tag{5.42}$$

In the same way by applying l'Hospital's rule twice we get

$$\lim_{\beta\to\infty} \beta^2 f(\beta^{-1}\boldsymbol{x}) = \lim_{\beta\to\infty} \frac{\frac{df(\beta^{-1}\boldsymbol{x})}{d\beta}}{\frac{d\beta^{-2}}{d\beta}} \tag{5.43}$$

$$= \lim_{\beta\to\infty} \frac{-\beta^{-2}(\nabla f(\beta^{-1}\boldsymbol{x}))^T\boldsymbol{x}}{-2\beta^{-3}} = \lim_{\beta\to\infty} \frac{(\nabla f(\beta^{-1}\boldsymbol{x}))^T\boldsymbol{x}}{2\beta^{-1}}$$

$$= \lim_{\beta\to\infty} \frac{\boldsymbol{x}^T\boldsymbol{H}_f(\beta^{-1}\boldsymbol{x})\boldsymbol{x}\beta^{-2}}{2\beta^{-2}} = \lim_{\beta\to\infty} \frac{1}{2}\boldsymbol{x}^T\boldsymbol{H}_f(\beta^{-1}\boldsymbol{x})\boldsymbol{x} = \frac{1}{2}\boldsymbol{x}\boldsymbol{H}_f(\mathrm{o})\boldsymbol{x}.$$

This gives for the integral over $h_\beta(\boldsymbol{x})$ then

$$\lim_{\beta\to\infty} \int_{\mathbb{R}^n} h_\beta(\boldsymbol{x})\, d\boldsymbol{x} = \int_{\mathbb{R}^n} g_0(o)|(\nabla g_1(\mathrm{o}))^T\boldsymbol{x}| \exp\left(\frac{1}{2}\boldsymbol{x}\boldsymbol{H}_f(\mathrm{o})\boldsymbol{x}\right)\, d\boldsymbol{x}. \tag{5.44}$$

Using theorem 26, p. 30 we get the final result

$$\lim_{\beta\to\infty} \int_{\mathbb{R}^n} h_\beta(\boldsymbol{x})\, d\boldsymbol{x} = 2(2\pi)^{(n-1)/2} g_0(\mathrm{o}) \left|\frac{(\nabla g_1(\mathrm{o}))^T \boldsymbol{H}_f^{-1}(\mathrm{o})\nabla g_1(\mathrm{o})}{\det(\boldsymbol{H}_f(\mathrm{o}))}\right|^{1/2}. \tag{5.45}$$

This gives the result. If F is not convex, we take a subset $F_K \subset F$, which is convex has the origin in its interior. It can be shown easily using lemma 39, p. 55 that the integral over $F \setminus F_K$ is negligible compared with the integral over F_K. □

We derive now an approximation if the function $h(\boldsymbol{x})$ vanishes at the maximum point of f or has a slight singularity. This is a modification of the approximation by Fulks/Sather [66]. This result contains the last as a special case.

Theorem 43 *Given is a compact set $F \subset \mathbb{R}^n$, a twice continuously differentiable function $f: \mathbb{R}^n \to \mathbb{R}$ and a continuous function $h_0: S_n(1) \to \mathbb{R}$ with $S_n(1)$ the n-dimensional unit sphere. Further for a point $\boldsymbol{x}_0$ in the interior of F let be defined a continuous function $h: F\setminus\{\boldsymbol{x}_0\} \to \mathbb{R}$, which has as $\boldsymbol{x} \to \boldsymbol{x}_0$ the form*

$$h(\boldsymbol{x}) \sim |\boldsymbol{x}-\boldsymbol{x}_0|^\nu h_0(|\boldsymbol{x}-\boldsymbol{x}_0|^{-1}(\boldsymbol{x}-\boldsymbol{x}_0)) + o(|\boldsymbol{x}-\boldsymbol{x}_0|^\nu) \tag{5.46}$$

with $\nu > -n$.

Given is the integral

$$I(\beta) = \int_F h(\boldsymbol{x}) \exp\left(\beta^2 f(\boldsymbol{x})\right)\, d\boldsymbol{x}. \tag{5.47}$$

If the function f achieves its global maximum with respect to F only at the point $\boldsymbol{x}_0$ and if the Hessian $\boldsymbol{H}_f(\boldsymbol{x}_0)$ is negative definite, the following asymptotic equation is valid as $\beta \to \infty$

$$I(\beta) \sim 2^{(n+\nu)/2-1}\Gamma\left(\frac{n+\nu}{2}\right)\frac{I_0}{|\det(\boldsymbol{H}_f(\boldsymbol{x}_0))|^{1/2}} \cdot e^{\beta^2 f(\boldsymbol{x}^*)}\beta^{-(n+\nu)}. \tag{5.48}$$

Here the constant I_0 is defined by a surface integral

$$I_0 = \int_{S_n(1)} |\boldsymbol{z}^T \boldsymbol{H}_f^{-1}(\boldsymbol{x}_0)\boldsymbol{z}|^{\nu/2} h_0(\boldsymbol{z})\, ds_n(\boldsymbol{z}) \tag{5.49}$$

over $S_n(1)$.

PROOF: We assume that $\boldsymbol{x}_0 = \mathbf{o}$ and $f(\mathbf{o}) = 0$. Further we assume that the eigenvectors of the Hessian $\boldsymbol{H}_f(\mathbf{o})$ are the n unit vectors $\boldsymbol{e}_1, \ldots, \boldsymbol{e}_n$ with negative eigenvalues $\lambda_1, \ldots, \lambda_n$.

We take an $\epsilon > 0$ such that the set $K_\epsilon = \{\boldsymbol{x}; |\boldsymbol{x}| \leq \epsilon\}$ is a subset of F and the Hessian $H_f(\boldsymbol{x})$ is negative definite for all $\boldsymbol{x} \in K_\epsilon$. We split the integral into two integrals

$$\int_F h(\boldsymbol{x}) \exp(\beta^2 f(\boldsymbol{x}))\, d\boldsymbol{x} \tag{5.50}$$

$$= \underbrace{\int_{K_\epsilon} h(\boldsymbol{x}) \exp(\beta^2 f(\boldsymbol{x}))\, d\boldsymbol{x}}_{=I_1(\beta)} + \underbrace{\int_{F\setminus K_\epsilon} h(\boldsymbol{x}) \exp(\beta^2 f(\boldsymbol{x}))\, d\boldsymbol{x}}_{=I_2(\beta)}.$$

For the second integral we get with $K_0 = \max_{\boldsymbol{x}\in F} |h(\boldsymbol{x})|$ that

$$|I_2(\beta)| \leq K_0 \int_{F\setminus K_\epsilon} \exp(\beta^2 f(\boldsymbol{x}))\, d\boldsymbol{x}. \tag{5.51}$$

Using lemma 40, p. 55 gives that there is a constant $k > 0$ with $f(\boldsymbol{x}) \leq -k|\boldsymbol{x}|^2$ and therefore

$$|I_2(\beta)| \leq K_0 \int_{F\setminus K_\epsilon} \exp(-\beta^2 k |\boldsymbol{x}|^2)\, d\boldsymbol{x} = o(\beta^r) \tag{5.52}$$

for all $r \in \mathbb{R}$.

For the first integral we use corollary 11, p. 13. We have a diffeomorphism $K_\epsilon \to V, \boldsymbol{x} \to \boldsymbol{y} = \boldsymbol{T}(\boldsymbol{x})$ as defined in the corollary such that $f(\boldsymbol{x}) = -|\boldsymbol{y}|^2/2$. Then we have

$$I_1(\beta) = \int_V |\boldsymbol{T}^{-1}(\boldsymbol{y})|^\nu [h_0(|\boldsymbol{y}|^{-1}\boldsymbol{y}) + o(|\boldsymbol{T}^{-1}(\boldsymbol{y})|^\nu)] \exp(-\beta^2|\boldsymbol{y}|^2/2) J(\boldsymbol{y}) d\boldsymbol{y}. \tag{5.53}$$

Here $J(\boldsymbol{y})$ is the Jacobian of the transformation $\boldsymbol{T}$. Since $\boldsymbol{y}$ points in the same direction as $\boldsymbol{x}$, we have that $h_0(|\boldsymbol{y}|^{-1}\boldsymbol{y}) = h_0(|\boldsymbol{x}|^{-1}\boldsymbol{x})$.

We take now a subset of V, a sphere with radius ϵ_1 around the origin. Using the same arguments as before, it is easy to show that we have asymptotically

$$I_1(\beta) \sim \int_{S_{\epsilon_1}} |T^{-1}(y)|^{\nu} h_0(|y|^{-1}y) \exp(-\beta^2|y|^2/2) J(y)\, dy, \; \beta \to \infty. \quad (5.54)$$

If we make a transformation to spherical coordinates (ρ, τ) with $\rho = |y|$ and $\tau = |y|^{-1}y$, we get for this integral

$$\sim \int_0^{\epsilon_1} \underbrace{\left[\int_{S_n(1)} |T^{-1}(\rho\tau)|^{\nu} h_0(\tau) J(\rho\tau)\, ds_n(\tau) \right]}_{=h^*(\rho)} \rho^{n-1} \exp(-\beta^2\rho^2/2)\, d\rho. \quad (5.55)$$

Since as $\rho \to 0$ we have $J(\rho\tau) \to |\det(H_f(o))|^{-1/2}$ and $T^{-1}(\rho\tau) \sim -\rho \sum_{i=1}^{n} \lambda_i^{-1/2} \tau_i$, we obtain for the function $h^*(\rho)$ the asymptotic form

$$h^*(\rho) \sim \rho^{\nu} I_0 |\det(H_f(o))|^{-1/2}, \; \rho \to 0. \quad (5.56)$$

Then we have for the integrand in the last equation the form

$$I_0 \rho^{\nu+n-1} \exp(-\beta^2\rho^2/2) |\det(H_f(o))|^{-1/2}. \quad (5.57)$$

Now, we can use theorem 36, p. 48 with $s = n + \nu$, $r = 2$, $a = 1$ and $b = I_0$ and we find asymptotically

$$I_1(\beta) \sim \frac{I_0}{2} \Gamma\left(\frac{n+\nu}{2}\right) 2^{(n+\nu)/2} \beta^{-(n+\nu)} |\det(H_f(o))|^{-1/2}, \; \beta \to \infty. \quad (5.58)$$

This gives the result in this case. The general case can be proved by making a linear transformation, which reduces it to this case. □

The next theorem is a generalization of theorem 41, p. 56. Here in the exponent additional to the function f which is multiplied by β^2 there is a function f_1 multiplied by β (see [25]). Such integrals appear if we consider moment generating functions of normal random variables.

Theorem 44 *Let be given:*

1. *A compact set $F \subset \mathbb{R}^n$ with the origin in its interior.*
2. *A twice continuously differentiable function $f : F \to \mathbb{R}$ with $f(x) < f(o)$ for all $x \neq o$. The Hessian $H_f(o)$ of f at o is negative definite.*
3. *A continuously differentiable function $f_1 : \mathbb{R}^n \to \mathbb{R}$.*
4. *A continuous function $h : F \to \mathbb{R}$ with $h(o) \neq 0$.*

For the integral

$$I(\beta) = \int_F h(\boldsymbol{x}) \exp(\beta f_1(\boldsymbol{x}) + \beta^2 f(\boldsymbol{x}))\, d\boldsymbol{x} \tag{5.59}$$

we have under these conditions the following asymptotic expression

$$I(\beta) = \int_F h(\boldsymbol{x}) \exp(\beta f_1(\boldsymbol{x}) + \beta^2 f(\boldsymbol{x}))\, d\boldsymbol{x} \tag{5.60}$$

$$\sim (2\pi)^{n/2} h(\mathbf{o}) \frac{\exp(-\frac{1}{2}(\nabla f_1(\mathbf{o}))^T \boldsymbol{H}_f^{-1}(\mathbf{o}) \nabla f_1(\mathbf{o}))}{\sqrt{|\det(\boldsymbol{H}_f(\mathbf{o}))|}} \frac{e^{\beta f_1(\mathbf{o}) + \beta^2 f(\mathbf{o})}}{\beta^n}, \quad \beta \to \infty.$$

PROOF: We assume for simplicity that $f(\mathbf{o}) = f_1(\mathbf{o}) = 0$. Further we assume first that F is convex. Define for all $\boldsymbol{x} \in I\!R^n \setminus F$ the functions f, f_1 and h by $f(\boldsymbol{x}) = f_1(\boldsymbol{x}) = h(\boldsymbol{x}) = 0$. Then these functions are defined on $I\!R^n$ and we can write the integral $I(\beta)$ in the form

$$I(\beta) = \int_{I\!R^n} h(\boldsymbol{x}) \exp\left(\beta f_1(\boldsymbol{x}) + \beta^2 f(\boldsymbol{x})\right)\, d\boldsymbol{x}. \tag{5.61}$$

Making the substitution $\boldsymbol{x} \to \boldsymbol{y} = \beta \boldsymbol{x}$ gives then

$$I(\beta) = \beta^{-n} \int_{I\!R^n} h(\beta^{-1}\boldsymbol{y}) \exp\left(\beta f_1(\beta^{-1}\boldsymbol{y}) + \beta^2 f(\beta^{-1}\boldsymbol{y})\right)\, d\boldsymbol{y}. \tag{5.62}$$

Let be defined the functions $g_\beta(\boldsymbol{x})$ by

$$g_\beta(\boldsymbol{y}) = h(\beta^{-1}\boldsymbol{y}) \exp\left(\beta f_1(\beta^{-1}\boldsymbol{y}) + \beta^2 f(\beta^{-1}\boldsymbol{y})\right) \tag{5.63}$$

for $\boldsymbol{x} \in F$ and zero elsewhere.

These functions are bounded by

$$|g_\beta(\boldsymbol{y})| \le k_0 \exp\left(\beta f_1(\beta^{-1}\boldsymbol{y}) + \beta^2 f(\beta^{-1}\boldsymbol{y})\right). \tag{5.64}$$

Here $k_0 = \max_{\boldsymbol{x} \in F} |h(\boldsymbol{x})|$. From lemma 40, p. 55 follows that there is a constant $k_1 > 0$ such that for vectors $\boldsymbol{y} \in \beta F$ always

$$\beta^2 f(\beta^{-1}\boldsymbol{y}) \le -k_1 |\boldsymbol{y}|^2. \tag{5.65}$$

If we define $k_2 = \sup_{\boldsymbol{x} \in F} |\nabla f_1(\boldsymbol{x})|$, we get (since F is convex) for all $\boldsymbol{y} \in \beta F$ further

$$|\beta f_1(\beta^{-1}\boldsymbol{y})| \le k_2 |\boldsymbol{y}|. \tag{5.66}$$

The last three inequalities give the estimate

$$|g_\beta(\boldsymbol{y})| \le k_0 \exp(k_2|\boldsymbol{y}| - k_1|\boldsymbol{y}|^2). \tag{5.67}$$

The function on the righthand side of the inequality is integrable over $\mathbb{R}^n$ as can be shown easily using spherical coordinates. This gives an integrable upper bound for the functions $|g_\beta|$ and so we can use theorem 8, p. 12. This yields

$$\lim_{\beta\to\infty} \int_{\mathbb{R}^n} g_\beta(\boldsymbol{y})\, d\boldsymbol{y} = \int_{\mathbb{R}^n} \lim_{\beta\to\infty} g_\beta(\boldsymbol{y})\, d\boldsymbol{y}. \tag{5.68}$$

We have to find the limit $\lim_{\beta\to\infty} g_\beta(\boldsymbol{y})$. Since h is continuous, we find

$$\lim_{\beta\to\infty} h(\beta^{-1}\boldsymbol{y}) = h(\mathbf{o}). \tag{5.69}$$

The limit of the product $\beta \cdot f_1(\beta^{-1}\boldsymbol{y})$ is obtained using l'Hospital's rule

$$\begin{aligned} \beta \cdot f_1(\beta^{-1}\boldsymbol{y}) = \frac{f_1(\beta^{-1}\boldsymbol{y})}{\beta^{-1}} &= \lim_{\beta\to\infty} \frac{\frac{df_1(\beta^{-1}\boldsymbol{y})}{d\beta}}{\frac{d\beta^{-1}}{d\beta}} \\ = \lim_{\beta\to\infty} \frac{-\beta^{-2}(\nabla f_1(\beta^{-1}\boldsymbol{y}))^T\boldsymbol{y}}{-\beta^{-2}} &= (\nabla f_1(\mathbf{o}))^T\boldsymbol{y}. \end{aligned} \tag{5.70}$$

Applying l'Hospital's rule twice gives further

$$\begin{aligned} \lim_{\beta\to\infty} \frac{f(\beta^{-1}\boldsymbol{y})}{\beta^{-2}} &= \lim_{\beta\to\infty} \frac{-\beta^{-2}(\nabla f(\beta^{-1}\boldsymbol{y}))^T\boldsymbol{y}}{-2\beta^{-3}} \\ = \lim_{\beta\to\infty} \frac{(\nabla f(\beta^{-1}\boldsymbol{y}))^T\boldsymbol{y}}{2\beta^{-1}} &= \lim_{\beta\to\infty} \frac{-\beta^{-2}\boldsymbol{y}^T\boldsymbol{H}_f(\beta^{-1}\boldsymbol{y})\boldsymbol{y}}{-2\beta^{-2}} \\ &= \frac{1}{2}\boldsymbol{y}^T\boldsymbol{H}_f(\mathbf{o})\boldsymbol{y}. \end{aligned} \tag{5.71}$$

From this we get then

$$\lim_{\beta\to\infty} g_\beta(\boldsymbol{y}) = h(\mathbf{o}) \exp\left((\nabla f_1(\mathbf{o}))^T\boldsymbol{y} + \frac{1}{2}\boldsymbol{y}^T\boldsymbol{H}_f(\mathbf{o})\boldsymbol{y}\right). \tag{5.72}$$

Applying now the Lebesgue theorem yields

$$\lim_{\beta\to\infty} \int_{\mathbb{R}^n} g_\beta(\boldsymbol{y})d\boldsymbol{y} = h(\mathbf{o}) \int_{\mathbb{R}^n} \exp\left((\nabla f_1(\mathbf{o}))^T\boldsymbol{y} + \frac{1}{2}\boldsymbol{y}^T\boldsymbol{H}_f(\mathbf{o})\boldsymbol{y}\right) d\boldsymbol{y} \tag{5.73}$$

and for this integral we find (see lemma 26, p. 30)

$$= (2\pi)^{n/2} h(\mathbf{o}) \frac{\exp(-\frac{1}{2}(\nabla f_1(\mathbf{o}))^T\boldsymbol{H}_f^{-1}(\mathbf{o})\nabla f_1(\mathbf{o}))}{\sqrt{|\det(\boldsymbol{H}_f(\mathbf{o}))|}}. \tag{5.74}$$

This proves the theorem for a convex set F.

If F is not convex, we take a convex subset F_c of F such that the origin lies in its interior. For this set we can prove the result. Then we have to show only that the integral over $F \setminus F_c$ is negligible in comparison with the integral over F_c, which can be done easily. □

5.4 Boundary Maxima

In this subsection some new results about asymptotic approximations for integrals are proved.

In the next lemma we give a bound for the decrease of a function near a regular maximum under constraints. This lemma is similar to lemma 40, p. 55.

If a function has the origin a maximum with respect to the set $\{\boldsymbol{x}; x_n \geq 0\}$ then due to the Lagrange multiplier rule the gradient is parallel to the x_n-axis and the matrix of the second derivatives with respect to the variables $x_1, \ldots, x_{n-1}$ must be negative semi-definite. If the gradient does not vanish and the matrix is negative definite, we can prove the following.

Lemma 45 *Let be given a twice continuously differentiable function $f : F \to \mathbb{R}$ with $F \subset \mathbb{R}^n$ such that*

1. *F is compact with the origin in its interior.*
2. *f has its global maximum with respect to $F^* = F \cap \{\boldsymbol{x}; x_n \geq 0\}$ at $\mathbf{o}$.*
3. *The gradient of f at $\mathbf{o}$ does not vanish, i.e. $\nabla f(\mathbf{o}) \neq \mathbf{o}$.*
4. *The $(n-1)\times(n-1)$ matrix $\boldsymbol{H}^*(\mathbf{o}) = (f^{ij}(\mathbf{o}))_{i,j=1,\ldots,n-1}$ is negative definite at $\mathbf{o}$.*

Then there exists a constant $k > 0$ such that for $\boldsymbol{x} \in F^*$ with $\boldsymbol{x} \neq \mathbf{o}$ always

$$f(\boldsymbol{x}) \leq f(\mathbf{o}) - k(|\tilde{\boldsymbol{x}}|^2 + |x_n|) \tag{5.75}$$

where $\tilde{\boldsymbol{x}} = (x_1, \ldots, x_{n-1})$.

PROOF: We assume that $f(\mathbf{o}) = 0$. We assume that no such constant exists. Then there must be a sequence $(\boldsymbol{x}^{(m)})$ of points in F^* such that

$$-m^{-1} < f(\boldsymbol{x}^{(m)})(|\tilde{\boldsymbol{x}}^{(m)}|^2 + |x_n^{(m)}|)^{-1} \leq 0. \tag{5.76}$$

Here we set $\tilde{\boldsymbol{x}}^{(m)} = (x_1^{(m)}, \ldots, x_{n-1}^{(m)})$.

Since F^* is compact there is a convergent subsequence and we assume without loss of generality that this subsequence is just the sequence itself. Since $f(\boldsymbol{x}^{(m)} \to 0$ we have $\boldsymbol{x}^{(m)} \to \mathbf{o}$. We split the difference

$$f(\boldsymbol{x}^{(m)}) = \left[f(\boldsymbol{x}^{(m)}) - f(\tilde{\boldsymbol{x}}^{(m)}, 0)\right] + \left[f(\tilde{\boldsymbol{x}}^{(m)}, 0) - f(0, \ldots, 0)\right]. \tag{5.77}$$

Making now a second order Taylor expansion of f with respect to $x_1, \ldots, x_{n-1}$ at the origin and a first Taylor expansion with respect to x_n at $(\tilde{x}^{(m)}, 0)$ to estimate the differences gives

$$\begin{aligned} & f(\tilde{x}^{(m)}, 0) - f(0, \ldots, 0) && (5.78) \\ = & \sum_{i=1}^{n-1} f^i(o) x_i^{(m)} + \frac{1}{2} (\tilde{x}^{(m)})^T H^*((\vartheta_1 \tilde{x}^{(m)}, 0)) \tilde{x}^{(m)} \\ = & \frac{1}{2} (\tilde{x}^{(m)})^T H^*((\vartheta_1 \tilde{x}^{(m)}, 0)) \tilde{x}^{(m)} \end{aligned}$$

and for the other difference using the mean value theorem

$$f(x^{(m)}) - f((\tilde{x}^{(m)}, 0)) = f^n((\tilde{x}^{(m)}, \vartheta_2 x_n^{(m)})) x_n^{(m)} \tag{5.79}$$

with $0 \le \vartheta_1, \vartheta_2 \le 1$.

Now, since $x^{(m)} \to o$, for m large enough we get

$$\begin{aligned} (\tilde{x}^{(m)})^T H^*((\vartheta_1 (x_m^*, 0))) \tilde{x}^{(m)} & \le \frac{1}{2} (\tilde{x}^{(m)})^T H^*(o) \tilde{x}^{(m)} && (5.80) \\ f^n((\tilde{x}^{(m)}, \vartheta_2 x_n^{(m)})) x_n^{(m)} & \le \frac{1}{2} f^n(o) x_n^{(m)} \end{aligned}$$

and so we obtain

$$f(x^{(m)}) \le \frac{1}{2} \left(f^n(o) x_n^{(m)} + \frac{1}{2} (\tilde{x}^{(m)})^T H^*(o) \tilde{x}^{(m)} \right). \tag{5.81}$$

Since $f^n(o) < 0$ and $H^*(o)$ is negative definite we get with $\lambda_0 < 0$ being the largest eigenvalue of this matrix that for m large enough that there is a constant $k_0 > 0$ such that

$$f(x^{(m)}) \le -\frac{1}{2} k_0 (|x_n^{(m)}| + |\tilde{x}^{(m)}|^2). \tag{5.82}$$

But this contradicts the properties of the chosen sequence. Therefore such a constant k must exist. □

In many of the theorems in this section we will assume that the following condition is fulfilled.

Condition A:

Given is a twice continuously differentiable function $g : \mathbb{R}^n \to \mathbb{R}$ such that by $F = \{x; g(x) \le 0\}$ is defined a compact set and by $G = \{x; g(x) = 0\}$ a compact C^2 hypersurface. The gradient $\nabla g(x)$ does not vanish on G and the surface G is oriented by the normal field $n(x) = |\nabla g(x)|^{-1} \nabla g(x)$.

In a number of the next results we have to compute the determinant of an $(n-1) \times (n-1)$-matrix $A^T H A$ with H an $n \times n$-matrix and $A = (a_1, \ldots, a_{n-1})$ an $n \times (n-1)$-matrix, where the a_i's are orthonormal vectors. Let a_n be the unit vector, which is orthogonal to $a_1, \ldots, a_{n-1}$.

We can use then corollary 19, p. 21 to simplify this. We have that

$$\det(A^T H A) = \det((I_n - P)^T H (I_n - P) + P^T P) \tag{5.83}$$

with $P = a_n a_n^T$ the projection matrix onto the subspace spanned by a_n.

In the following theorem the case is treated that the global maximum of the function is on the boundary of the integration domain and that the function g defining the boundary is smooth at the maximum point. In this case the gradient of the function f does not vanish, it is parallel to the normal vector of the surface at this point, i.e. the first partial derivatives in all directions orthogonal to this are zero.

Theorem 46 *Let condition A be fulfilled. Further are given a twice continuously differentiable function $f : \mathbb{R}^n \to \mathbb{R}$ and a continuous function $h : \mathbb{R}^n \to \mathbb{R}$.*

Assume that the following conditions are satisfied

1. *The function f attains its global maximum with respect to F only at the point $\boldsymbol{x}^* \in G$.*

2. *At $\boldsymbol{x}^*$ the gradient $\nabla f(\boldsymbol{x}^*)$ does not vanish.*

3. *In $\boldsymbol{x}^* \in M$ the $(n-1)\times(n-1)$-matrix $\boldsymbol{H}^*(\boldsymbol{x}^*)$ is regular. This matrix is defined by*
$$\boldsymbol{H}^*(\boldsymbol{x}^*) = \boldsymbol{A}^T(\boldsymbol{x}^*)\boldsymbol{H}(\boldsymbol{x}^*)\boldsymbol{A}(\boldsymbol{x}^*)$$
with
$$\boldsymbol{H}(\boldsymbol{x}^*) = \left(f^{ij}(\boldsymbol{x}^*) - \frac{|\nabla f(\boldsymbol{x}^*)|}{|\nabla g(\boldsymbol{x}^*)|} g^{ij}(\boldsymbol{x}^*)\right)_{i,j=1,\dots,n}$$
and
$$\boldsymbol{A}(\boldsymbol{x}^*) = (\boldsymbol{a}_1(\boldsymbol{x}^*), \dots, \boldsymbol{a}_{n-1}(\boldsymbol{x}^*)).$$
The $\boldsymbol{a}_1(\boldsymbol{x}^*), \dots, \boldsymbol{a}_{n-1}(\boldsymbol{x}^*)$ form an orthonormal basis of the tangential space of G at $\boldsymbol{x}^*$.

Then as $\beta \to \infty$ the following asymptotic equation is valid

$$\int_F h(\boldsymbol{x})e^{\beta^2 f(\boldsymbol{x})}d\boldsymbol{x} \sim (2\pi)^{(n-1)/2}\frac{h(\boldsymbol{x}^*)}{\sqrt{|(\nabla f(\boldsymbol{x}^*))^T \boldsymbol{C}(\boldsymbol{x}^*)\nabla f(\boldsymbol{x}^*)|}}\frac{e^{\beta^2 f(\boldsymbol{x}^*)}}{\beta^{n+1}} \tag{5.84}$$

$$= (2\pi)^{(n-1)/2}\frac{h(\boldsymbol{x}^*)}{|\nabla f(\boldsymbol{x}^*)|\sqrt{|\det(\boldsymbol{H}^*(\boldsymbol{x}^*))|}}\frac{e^{\beta^2 f(\boldsymbol{x}^*)}}{\beta^{n+1}}.$$

Here the $n \times n$-Matrix $\boldsymbol{C}(\boldsymbol{x}^*)$ is the cofactor matrix of the $n \times n$-Matrix $\boldsymbol{H}(\boldsymbol{x}^*)$.

PROOF: A complete proof is given in [15], p. 340 und [59], p. 82. The idea of the proof is similar to the one in the last theorem. We give two proofs. The first is an outline of a proof by making Taylor expansions and comparing the integrals of the Taylor approximation with the original integral.

We assume for simplicity that $\boldsymbol{x}^* = \mathbf{o}$, $f(\mathbf{o}) = 0$ and that at $\mathbf{o}$ the tangential space of G is spanned by the vectors $\boldsymbol{e}_1, \dots, \boldsymbol{e}_{n-1}$. That can always be achieved by a coordinate transformation and by adding a suitable constant. Since at $\mathbf{o}$ there is a maximum of f with respect to F and so also with respect to G, the Lagrange multiplier rule states that $f^i(\mathbf{o}) = 0$ for $i = 1, \dots, n-1$.

To simplify further we assume that the second derivatives of the function g at this point vanish, i.e. the main curvatures of the surface are all zero and therefore the surface G is not curved at this point. If the surface is curved at this point, we have only to replace in the final result the second derivatives of the function f by the corresponding second derivatives with respect to the local coordinates of the surface. These were calculated as in lemma 14, p. 17.

Outline of the first proof:
We write

$$I(\beta) = \int_F h(\boldsymbol{x}) e^{\beta^2 f(\boldsymbol{x})}\, d\boldsymbol{x}. \tag{5.85}$$

Now we approximate the function f near $\mathbf{o}$ by its second order Taylor expansion at $\mathbf{o}$. This gives

$$f(\boldsymbol{x}) \sim \frac{1}{2}\sum_{i,j=1}^{n-1} f^{ij}(\mathbf{o})x_i x_j + \frac{1}{2}\sum_{i=1}^{n} f^{in}(\mathbf{o})x_i x_n + f^n(\mathbf{o})x_n. \tag{5.86}$$

Replacing $h(\boldsymbol{x})$ by $h(\mathbf{o})$ we obtain

$$I(\beta) \sim h(\mathbf{o}) \int_F \exp\left[\frac{\beta^2}{2}\left(\sum_{i,j=1}^{n-1} f^{ij}(\mathbf{o})x_i x_j + \sum_{i=1}^{n} f^{in}(\mathbf{o})x_i x_n + 2f^n(\mathbf{o})x_n\right)\right] d\boldsymbol{x}.$$

By estimating the changes it can be shown again that the integration domain for the variables $x_1, \ldots, x_{n-1}$ can be replace by $\mathbb{R}^{n-1}$ and the domain for x_n by $[0, \infty)$ without changing the asymptotic behavior. This yields then

$$I(\beta) \sim h(\mathbf{o}) \int_{\mathbb{R}^{n-1}} \int_0^{\infty} \exp\left[\frac{\beta^2}{2}\left(\sum_{i,j=1}^{n-1} f^{ij}(\mathbf{o})x_i x_j \right.\right. \tag{5.87}$$

$$\left.\left. + \sum_{i=1}^{n} f^{in}(\mathbf{o})x_i x_n + 2f^n(\mathbf{o})x_n\right)\right] dx_n\, dx_1 \ldots dx_{n-1}$$

With the substitution $x_i \to u_i = \beta x_i$ for $i = 1, \ldots, n-1$ and $x_n \to u_n = \beta^2 x_n$ we get

$$= \frac{h(\mathbf{o})}{\beta^{n+1}} \int_{\mathbb{R}^{n-1}} \int_0^{\infty} \exp\left[\frac{1}{2}\left(\sum_{i,j=1}^{n-1} f^{ij}(\mathbf{o})u_i u_j \right.\right. \tag{5.88}$$

$$\left.\left. + \beta^{-1}\sum_{i=1}^{n} f^{in}(\mathbf{o})u_i u_n + 2f^n(\mathbf{o})u_n\right)\right] du_n \ldots du_1 \ldots du_{n-1}.$$

The term in the exponent divided by β is asymptotically negligible and therefore

$$I(\beta) \sim \frac{h(o)}{\beta^{n+1}} \int_{\mathbb{R}^{n-1}} \exp\left(\frac{1}{2}\sum_{i,j=1}^{n-1} f^{ij}(o)u_i u_j\right) du_1 \dots du_{n-1} \quad (5.89)$$
$$\times \int_0^\infty \exp(f^n(o)u_n)\, du_n.$$

For the first integral the result is (see lemma 26, p. 30)

$$\int_{\mathbb{R}^{n-1}} \exp\left(\frac{1}{2}\sum_{i,j=1}^{n-1} f^{ij}(o)u_i u_j\right) du_1 \dots du_{n-1} = \frac{(2\pi)^{(n-1)/2}}{|\det(\boldsymbol{H}^*(o))|^{1/2}}. \quad (5.90)$$

Here $\boldsymbol{H}^*(o) = (f^{ij}(o))_{i,j=1,\dots,n-1}$.

Since all components of the gradient of f at o with the exception of the n-th are equal zero, $|\nabla f(o)| = |f^n(o)|$ and this gives for the second integral

$$\int_0^\infty \exp(f^n(o)u_n)\, du_n = |\nabla f(o)|^{-1}. \quad (5.91)$$

The final result is then

$$\int_F h(\boldsymbol{x})e^{\beta^2 f(\boldsymbol{x})}\, d\boldsymbol{x} \sim (2\pi)^{(n-1)/2} \frac{h(o)}{|\nabla f(o)|\sqrt{|\det(\boldsymbol{H}^*(o))|}} \cdot \beta^{-(n+1)}. \quad (5.92)$$

Making a suitable rotation to change the coordinates it is always possible to transform the integral in such a form. Then the second derivatives transform as derived in lemma 7, p. 12. This gives then the general result.

Second proof:
We define the function $h_\beta(\boldsymbol{x})$ by

$$h_\beta(\boldsymbol{x}) = h(\beta^{-1}\tilde{\boldsymbol{x}}, \beta^{-2}x_n)\exp(\beta^2 f(\beta^{-1}\boldsymbol{x}, \beta^{-2}x_n)) \quad (5.93)$$

with $\tilde{\boldsymbol{x}} = (x_1, \dots, x_{n-1})$ for $\boldsymbol{x} \in F$ and zero elsewhere. We have then that

$$\beta^{n+1} I(\beta) = \int_{\mathbb{R}^n} h_\beta(\boldsymbol{x})\, d\boldsymbol{x}. \quad (5.94)$$

Using lemma 45, p. 65 we get that there is a constant $K_1 > 0$ with

$$|h_\beta(\boldsymbol{x})| \le K_1 \cdot \exp(-k(|\tilde{\boldsymbol{x}}|^2 + x_n)), \quad (5.95)$$

where $K_1 = \max_{\boldsymbol{x}\in F} |h(\boldsymbol{x})|$. But this function is integrable over the domain F and over $\mathbb{R}^n$ and therefore we can apply the Lebesgue theorem if we find $\lim_{\beta\to\infty} h_\beta(\boldsymbol{x})$. We get

$$\lim_{\beta\to\infty} h(\beta^{-1}\tilde{\boldsymbol{x}}, \beta^{-2}x_n) = h(o). \quad (5.96)$$

For the argument of the exponent we have

$$\lim_{\beta\to\infty} \beta^{-2} f(\beta^{-1}\tilde{x}, \beta^{-2}x_n) = \lim_{\beta\to\infty} \frac{f(\beta^{-1}\tilde{x}, \beta^{-2}x_n)}{\beta^{-2}}. \tag{5.97}$$

Applying now l'Hospital's rule once gives

$$\begin{aligned} &\lim_{\beta\to\infty} \frac{f(\beta^{-1}\tilde{x}, \beta^{-2}x_n)}{\beta^{-2}} \\ = &\lim_{\beta\to\infty} \frac{\beta^{-2}\sum_{i=1}^{n-1} f^i(\beta^{-1}\tilde{x}, \beta^{-2}x_n) + 2\beta^{-3} f^n(\beta^{-1}\tilde{x}, \beta^{-2}x_n)}{-2\beta^{-3}} \\ = &\lim_{\beta\to\infty} \frac{\sum_{i=1}^{n-1} f^i(\beta^{-1}\tilde{x}, \beta^{-2}x_n)}{2\beta^{-1}} + \lim_{\beta\to\infty} f^n(\beta^{-1}\tilde{x}, \beta^{-2}x_n) \\ = &\lim_{\beta\to\infty} \frac{\sum_{i=1}^{n-1} f^i(\beta^{-1}\tilde{x}, \beta^{-2}x_n)}{2\beta^{-1}} + f^n(o)x_n. \end{aligned} \tag{5.98}$$

To find the limit of the first summand we apply again the rule yielding then

$$\begin{aligned} &\lim_{\beta\to\infty} \frac{\sum_{i=1}^{n-1} f^i(\beta^{-1}\tilde{x}, \beta^{-2}x_n)}{2\beta^{-1}} \\ = &\lim_{\beta\to\infty} \frac{-\beta^{-2}\sum_{i,j=1}^{n-1} f^{ij}(\beta^{-1}\tilde{x}, \beta^{-2}x_n)x_i x_j + 2\beta^{-3}\sum_{i=1}^{n} f^{in}(\beta^{-1}\tilde{x}, \beta^{-2}x_n)x_i x_n}{-2\beta^{-2}} \\ = &\frac{1}{2}\lim_{\beta\to\infty} \sum_{i,j=1}^{n-1} f^{ij}(\beta^{-1}x, \beta^{-2}x_n)x_i x_j = \frac{1}{2}x^T H^*(o)x. \end{aligned}$$

□

This result can be proved in a modified form for surface integrals.

Theorem 47 *Let condition A be fulfilled. Further let be given a twice differentiable function $f : \mathbb{R}^n \to \mathbb{R}$ and a continuous function $h : \mathbb{R}^n \to \mathbb{R}$.*
Assume further that:

1. *The function f achieves its global maximum with respect to G only at the point $x^* \in G$.*

2. *At $x^* \in M$ the $(n-1)\times(n-1)$ matrix $H^*(x^*)$ is regular. This matrix is defined by*
$$H^*(x^*) = A^T(x^*)H(x^*)A(x^*)$$
with
$$H(x^*) = \left(f^{ij}(x^*) - \frac{|\nabla f(x^*)|}{|\nabla g(x^*)|}g^{ij}(x^*)\right)_{i,j=1,\ldots,n}$$
and
$$A(x^*) = (a_1(x^*), \ldots, a_{n-1}(x^*)).$$
Here the $a_1(x^*), \ldots, a_{n-1}(x^*)$ are an orthonormal basis of the tangential space of G at x^*.

Then as $\beta \to \infty$ the following asymptotic relation is valid

$$\int_G h(\boldsymbol{x})e^{\beta^2 f(\boldsymbol{x})}\, ds_G(\boldsymbol{x}) \sim (2\pi)^{(n-1)/2}\frac{h(\boldsymbol{x}^*)}{\sqrt{|\det(\boldsymbol{H}^*(\boldsymbol{x}^*))|}}e^{\beta^2 f(\boldsymbol{x}^*)}\beta^{-(n-1)}. \quad (5.99)$$

PROOF: Analogous to the proof of the last theorem. □

The following theorem is from the article [33] and treats the case that the maximum point is in the intersection of several manifolds.

Theorem 48 *Given are:*

1. *A twice continuously differentiable function $f : \mathbb{R}^n \to \mathbb{R}$.*
2. *m twice continuously differentiable functions $g_i : \mathbb{R}^n \to \mathbb{R}$.*
3. *A continuous function $h : \mathbb{R}^n \to \mathbb{R}$.*

The functions $g_1, \ldots, g_m$ define a compact set $F = \cap_{i=1}^m \{\boldsymbol{x}; g_i(\boldsymbol{x}) \le 0\}$.
Further the following conditions are fulfilled:

1. *The function f achieves its global maximum with respect to F only at the point $\boldsymbol{x}^*$.*
2. *There is a $k \in \{1, \ldots, m\}$ such that*

$$g_i(\boldsymbol{x}^*) = 0 \text{ for } i = 1, \ldots, k \text{ and}$$
$$g_j(\boldsymbol{x}^*) < 0 \text{ for } j = k+1, \ldots, m.$$

3. *The gradients $\nabla g_i(\boldsymbol{x}^*)$ with $i = 1, \ldots, k$ are linearly independent.*
4. *The gradient $\nabla f(\boldsymbol{x}^*)$ has a unique representation in the form*

$$\nabla f(\boldsymbol{x}^*) = \sum_{i=1}^k \gamma_i \nabla g_i(\boldsymbol{x}^*)$$

with $\gamma_i > 0$ for $i = 1, \ldots, k$. (From 3 follows with theorem 15, p. 19 that the gradient has such a representation with $\gamma_i \ge 0$.)

5. *The matrix $\boldsymbol{H}^*(\boldsymbol{x}^*)$ is regular. Here*

$$\boldsymbol{H}^*(\boldsymbol{x}^*) = \boldsymbol{A}^T(\boldsymbol{x}^*)\boldsymbol{H}(\boldsymbol{x}^*)\boldsymbol{A}(\boldsymbol{x}^*)$$

with

$$\boldsymbol{H}(\boldsymbol{x}^*) = (f^{ij}(\boldsymbol{x}^*) - \sum_{r=1}^k \gamma_r g_r^{ij}(\boldsymbol{x}^*))_{i,j=1,\ldots,n}$$

and

$$\boldsymbol{A}(\boldsymbol{x}^*) = (\boldsymbol{a}_1(\boldsymbol{x}^*), \ldots, \boldsymbol{a}_{n-k}(\boldsymbol{x}^*)).$$

The $\boldsymbol{a}_1(\boldsymbol{x}^*), \ldots, \boldsymbol{a}_{n-k}(\boldsymbol{x}^*)$ are an orthonormal basis of the subspace of $\mathbb{R}^n$ which it orthogonal to the subspace $\text{span}[\nabla g_1(\boldsymbol{x}^*), \ldots, \nabla g_k(\boldsymbol{x}^*)]$.

Then the asymptotic relation holds

$$\int_F h(\boldsymbol{x}) e^{\beta^2 f(\boldsymbol{x})}\, d\boldsymbol{x} \tag{5.100}$$

$$\sim \quad (2\pi)^{(n-k)/2} \frac{h(\boldsymbol{x}^*)}{\prod_{i=1}^k \gamma_i \sqrt{\det(\boldsymbol{G}) \cdot |\det(\boldsymbol{H}^*(\boldsymbol{x}^*))|}} \cdot e^{\beta^2 f(\boldsymbol{x}^*)} \beta^{-(n+k)}, \ \beta \to \infty.$$

Here $\boldsymbol{G} = ((\nabla g_i(\boldsymbol{x}^*))^T \cdot \nabla g_j(\boldsymbol{x}^*))_{i,j=1,\ldots,k}$ *is the Gramian of the vectors* $\nabla g_1(\boldsymbol{x}^*), \ldots, \nabla g_k(\boldsymbol{x}^*)$.

PROOF: We give only a short outline of the proof. A complete proof can be found in [33]. We assume that the maximum point $\boldsymbol{x}^*$ is the origin $\mathbf{o}$, that $f(\mathbf{o}) = 0$ and that the subspace spanned by the k column vectors of $\boldsymbol{A}$ is equal to the subspace spanned by the first k unit vectors $\boldsymbol{e}_i, i = 1, \ldots, k$. Then the subspace orthogonal to this space is spanned by the unit vectors $\boldsymbol{e}_{k+1}, \ldots, \boldsymbol{e}_n$.

In [33] it is shown that the integrand can be replaced near the maximum point by its second order Taylor expansion and the function $h(\boldsymbol{x})$ by its value at this point without changing the asymptotic behavior of the integrals as $\beta \to \infty$. We can write therefore

$$I(\beta) \sim h(\mathbf{o}) \int_F \exp\left[\beta^2 \left((\nabla f(\mathbf{o}))^T \boldsymbol{x} + \frac{1}{2}\boldsymbol{x}^T \boldsymbol{H}_f(\mathbf{o}) \boldsymbol{x}\right)\right] d\boldsymbol{x}. \tag{5.101}$$

If a coordinate change is made such that the vectors $-\nabla g_1(\mathbf{o}), \ldots, -\nabla g_k(\mathbf{o})$ become the new first k basis vectors instead of $\boldsymbol{e}_1, \ldots, \boldsymbol{e}_k$ and the other remain unchanged the transformation determinant is equal to the volume of the k-dimensional parallelepiped spanned by the k vectors $\nabla g_1(\mathbf{o}), \ldots, \nabla g_k(\mathbf{o})$ (see [62], p. 206-8). But this volume is the square root of the Gramian $\det(\boldsymbol{G}) = ((\nabla g_i(\mathbf{o}))^T \cdot \nabla g_j(\mathbf{o}))_{i,j=1,\ldots,k}$ (see lemma 3, p. 10), yielding

$$I(\beta) \sim \frac{h(\mathbf{o}) e^{\beta^2 f(\mathbf{o})}}{\sqrt{\det(\boldsymbol{G})}} \int_F \exp\left\{\beta^2 \left[-\sum_{i=1}^k \gamma_i u_i + \frac{1}{2}\left(\sum_{m,l=k+1}^n f^{ml}(\mathbf{o}) u_m u_l \right.\right.\right.$$
$$\left.\left.\left. + \sum_{j=1}^k \sum_{m=k+1}^n f^{jm}(\mathbf{o}) u_j u_m + \sum_{j=1}^k \sum_{m=1}^k f^{jm}(\mathbf{o}) u_j u_k \right)\right]\right\} d\boldsymbol{u}. \tag{5.102}$$

Here $\nabla f(\mathbf{o})$ is replaced by the linear combination $\sum_{i=1}^k \gamma_i \nabla g_i(\mathbf{o})$ and the derivatives are with respect to the new coordinates. But since only the first k coordinates are changed, the second derivatives, which involve only the coordinates $u_{k+1}, \ldots, u_n$ remain the same.

With the transformation $u_i \to w_i = \beta^2 u_i$ for $i = 1, .., k$ and $u_j \to w_j = \beta u_j$ for $j = k+1, .., n$ we get

$$I(\beta) \sim \beta^{-(n+k)} \frac{h(\mathrm{o}) e^{\beta^2 f(\mathrm{o})}}{\sqrt{\det(G)}} \int_{\tilde{F}} \exp\left\{ -\sum_{i=1}^{k} \gamma_i w_i + \frac{1}{2}\left[\sum_{m,l=k+1}^{n} f^{ml}(\mathrm{o}) w_m w_l \right.\right. \quad (5.103)$$

$$\left.\left. + \beta^{-1} \sum_{j=1}^{k} \sum_{m=k+1}^{n} f^{jm}(\mathrm{o}) w_j w_m + \beta^{-2} \sum_{j=1}^{k} \sum_{m=1}^{k} f^{jm}(\mathrm{o}) w_j w_m \right] \right\} dw.$$

$\tilde{F}$ is the transformed domain F.

As $\beta \to \infty$ the terms in the exponent multiplied by negative powers of β become negligible and we obtain then

$$I(\beta) \sim \frac{h(\mathrm{o}) e^{\beta^2 f(\mathrm{o})}}{\beta^{n+k} \sqrt{\det(G)}} \int_{\tilde{F}} \exp\left\{ -\sum_{i=1}^{k} \gamma_i w_i + \frac{1}{2} \sum_{m,l=k+1}^{n} f^{ml}(\mathrm{o}) w_m w_l \right\} dw. \quad (5.104)$$

This integral is now written as the product of two integrals. Introducing as new coordinates local parameters $v_{k+1}, \dots, v_n$ of the $(n-k)$-dimensional manifold $\cap_{i=1}^{k} \{x; g_i(x) = 0\}$. The integration domain is changed to $\mathbb{R}_+^k \times \mathbb{R}^{n-k}$, giving

$$I(\beta) \sim \beta^{-(n+k)} \frac{h(\mathrm{o}) e^{\beta^2 f(\mathrm{o})}}{\sqrt{\det(G)}} \int_{\mathbb{R}_+^k} \exp\left(-\sum_{i=1}^{k} \gamma_i w_i \right) dw_1 \cdots dw_k \quad (5.105)$$

$$\times \int_{\mathbb{R}^{n-k}} \exp\left(\frac{1}{2} \sum_{m,l=k+1}^{n} \tilde{f}^{lm}(\mathrm{o}) v_l v_k \right) D(v_{k+1}, \dots, v_n) dv_{k+1} \cdots dv_n.$$

Here $D(v_{k+1}, \dots, v_n)$ is the transformation determinant for the coordinate change to local coordinates. Without loss of generality we can choose these coordinates in such a way that $D(0, \dots, 0) = 1$. So we get finally

$$I(\beta) \sim (2\pi)^{(n-k)/2} \frac{h(\mathrm{o})}{\prod_{i=1}^{k} \gamma_i \sqrt{\det(G) \det(\tilde{f}^{lm}(\mathrm{o}))_{l,m=k+1,\dots,n}}} \frac{e^{\beta^2 f(\mathrm{o})}}{\beta^{n+k}}. \quad (5.106)$$

Due to the curvature of the manifold $\cap_{i=1}^{k} \{x; g_i(x) = 0\}$ at the point o we get additionally to the second derivatives of f a curvature factor appears (see lemma 14, p. 17) and the elements $\tilde{f}^{lm}(\mathrm{o})$ are then given by

$$\tilde{f}^{lm}(\mathrm{o}) = \frac{\partial^2 f(\mathrm{o})}{\partial x_l \partial x_m} - \sum_{i=1}^{k} \gamma_i \frac{\partial^2 g_i(\mathrm{o})}{\partial x_l \partial x_m}. \quad (5.107)$$

In the last equation only the determinant of the second derivatives depends on the coordinate system. By a suitable rotation the value of this determinant can be found

$$\det(\boldsymbol{H}^*(\mathbf{o})) = \det(\boldsymbol{A}(\mathbf{o})^T \boldsymbol{H}(\mathbf{o}) \boldsymbol{A}(\mathbf{o})). \tag{5.108}$$

This gives the result. □

In this case the asymptotic order of magnitude of the integral is $\exp(\beta^2 f(\boldsymbol{x}^*))\beta^{-(n+k)}$, it depends on the number k of active constraints in the point $\boldsymbol{x}^*$.

The following theorem treats the case that the maximum of the function is attained on a surface part and not only at one point.

Theorem 49 *Let condition A be fulfilled. Further is given a twice continuously differentiable functions $f : \mathbb{R}^n \to \mathbb{R}$ and a continuous function $h : \mathbb{R}^n \to \mathbb{R}$. Assume further:*

1. *The function f achieves its global maximum $m = \max_{\boldsymbol{x} \in F} f(\boldsymbol{x})$ with respect to F exactly on the compact k-dimensional submanifold $M \subset G$.*

2. *$\nabla f(\boldsymbol{x}) \neq \mathbf{o}$ for all $\boldsymbol{x} \in M$.*

3. *At all points $\boldsymbol{x} \in M$ the $(n-k-1) \times (n-k-1)$-matrix $\boldsymbol{H}^*(\boldsymbol{x})$ is regular. This matrix is defined by*

$$\boldsymbol{H}^*(\boldsymbol{x}) = \boldsymbol{A}^T(\boldsymbol{x}) \boldsymbol{H}(\boldsymbol{x}) \boldsymbol{A}(\boldsymbol{x})$$

with

$$\boldsymbol{H}(\boldsymbol{x}) = \left(f^{ij}(\boldsymbol{x}) - \frac{|\nabla f(\boldsymbol{x})|}{|\nabla g(\boldsymbol{x})|} g^{ij}(\boldsymbol{x}) \right)_{i,j=1,\ldots,n}$$

and

$$\boldsymbol{A}(\boldsymbol{x}) = (\boldsymbol{a}_1(\boldsymbol{x}), \ldots, \boldsymbol{a}_{n-k-1}(\boldsymbol{x})).$$

Here the $\boldsymbol{a}_1(\boldsymbol{x}), \ldots, \boldsymbol{a}_{n-k-1}(\boldsymbol{x})$ form an orthonormal basis of the subspace of the tangential space $T_G(\boldsymbol{x})$ which is orthogonal to $T_M(\boldsymbol{x})$.

Then the following asymptotic relation holds

$$\int_F h(\boldsymbol{x}) e^{\beta^2 f(\boldsymbol{x})}\, d\boldsymbol{x} \tag{5.109}$$

$$\sim (2\pi)^{(n-k-1)/2} \int_M \frac{h(\boldsymbol{x})\, ds_M(\boldsymbol{x})}{|\nabla f(\boldsymbol{x})| \sqrt{|\det(\boldsymbol{H}^*(\boldsymbol{x}))|}} \cdot e^{\beta^2 m} \beta^{-(n-k+1)}, \ \beta \to \infty.$$

Here $ds_M(\boldsymbol{x})$ denotes surface integration over M.

Proof: We can restrict the integration domain to a compact neighborhood V of M. We have then that

$$\int_F h(\boldsymbol{x}) e^{\beta^2 f(\boldsymbol{x})} d\boldsymbol{x} \sim \int_{F \cap V} h(\boldsymbol{x}) e^{\beta^2 f(\boldsymbol{x})} d\boldsymbol{x}, \ \beta \to \infty. \tag{5.110}$$

We assume that in the set $F \cap V$ exists a global coordinate system in the form $(h_1, \ldots, h_k, h_{k+1}, \ldots, h_{n-1}, g)$ with $(h_1, \ldots, h_k)$ being a global cordinate coordinate system for M. If no such system exists we have to make a partition of unity to find subsets where we can define local coordinate systems (see [130], p. 150).

The integral can now be written in the form

$$I(\beta) \sim \int_U h(T(\boldsymbol{u}^*, \tilde{\boldsymbol{u}}, u_n)) e^{\beta^2 f(T(\boldsymbol{u}^*, \tilde{\boldsymbol{u}}, u_n))} D(T(\boldsymbol{u}^*, \tilde{u}, u_n)) \, d\boldsymbol{u}^* d\tilde{\boldsymbol{u}} du_n. \quad (5.111)$$

Here $\boldsymbol{u}^* = (u_1, \ldots, u_k)$ and $\tilde{\boldsymbol{u}} = (u_{k+1}, \ldots, u_{n-1})$ and $u_n = -g(\boldsymbol{x})$. U is an open subset of $\mathbb{R}^n$ and $D(T(\boldsymbol{u}^*, \tilde{\boldsymbol{u}}, u_n))$ is the transformation determinant. The $u_1, \ldots, u_k$ are a parametrization of M.

We assume first that at a point $\boldsymbol{x}$ the tangential space $T_M(\boldsymbol{x})$ is spanned by the unit vectors $\boldsymbol{e}_1, \ldots, \boldsymbol{e}_k$ and the subspace of $T_G(\boldsymbol{x})$ orthogonal to this space by the unit vectors $\boldsymbol{e}_{k+1}, \ldots, \boldsymbol{e}_{n-1}$. Then the gradient of f at this point is parallel to the unit vector $\boldsymbol{e}_n$ due to the Lagrange multiplier rule, since here is a maximum with respect to G.

We write the integral in the form

$$I(\beta) \sim \int_{U^*} \left[\underbrace{\int_{\tilde{U}} \int_{-\delta}^{0} h(T(\boldsymbol{u}^*, \tilde{\boldsymbol{u}}, u_n)) e^{\beta^2 f(T(\boldsymbol{u}^*, \tilde{\boldsymbol{u}}, u_n))} D(T(\boldsymbol{u}^*, \tilde{u}, u_n)) d\tilde{\boldsymbol{u}} du_n}_{=I(\boldsymbol{u}^*, \beta)} \right] d\boldsymbol{u}^*.$$

We consider now the integral in square brackets. For it we can for all points in $\boldsymbol{x} \in M$ derive an asymptotic approximation

$$I(\boldsymbol{u}^*, \beta) = \int_{\tilde{U}} \int_{-\delta}^{0} h(T(\boldsymbol{u}^*, \tilde{\boldsymbol{u}}, u_n)) e^{\beta^2 f(T(\boldsymbol{u}^*, \tilde{\boldsymbol{u}}, u_n))} D(T(\boldsymbol{u}^*, \tilde{u}, u_n)) du_n d\tilde{\boldsymbol{u}}. \quad (5.112)$$

Without loss of generality we assume always that $D(T(\boldsymbol{u}^*, 0, 0)) = 1$ for the point we consider. If this is not the case it can be achieved by a suitable scaling of the local coordinates.

For a given $\boldsymbol{u}^*$ the function $f(T(\boldsymbol{u}^*, \tilde{\boldsymbol{u}}, u_n))$ has a global maximum at $f(T(\boldsymbol{u}^*, \mathrm{o}, 0))$ with respect to $\tilde{\boldsymbol{U}} \times [-\delta, 0]$.

We compute now the second partial derivatives of this function with respect to the variables $u_{k+1}, \ldots, u_{n-1}$ and the first derivative with respect to u_n. The second derivatives are found using lemma 14, p. 17

$$\frac{\partial f(T(\boldsymbol{u}^*, \tilde{\boldsymbol{u}}, u_n))}{\partial u_n} = f^n(\boldsymbol{x}), \quad (5.113)$$

$$\frac{\partial^2 f(T(\boldsymbol{u}^*, \tilde{\boldsymbol{u}}, u_n))}{\partial u_i u_j} = f^{ij}(\boldsymbol{x}) - \frac{|\nabla f(\boldsymbol{x})|}{|\nabla g(\boldsymbol{x})|} g^{ij}(\boldsymbol{x}), \; i, j = k+1, \ldots, n-1$$

with $\boldsymbol{x} = T(\boldsymbol{u}^*, \tilde{\boldsymbol{u}}, u_n)$.

Since f has a local maximum in all points of M with respect to G, its gradient is parallel at all these points to $\nabla g(\boldsymbol{x})$. From this follows

$$|f^n(\boldsymbol{x})| = |\nabla f(\boldsymbol{x})|. \tag{5.114}$$

So we get for the integral in square brackets with theorem 46, p. 67 the following asymptotic approximation

$$I(\boldsymbol{u}^*, \beta) \sim \frac{(2\pi)^{(n-k-1)/2} h(\boldsymbol{x})}{|\nabla f(\boldsymbol{x})| \sqrt{|\det(\boldsymbol{H}^*(\boldsymbol{x}))|}} \frac{e^{\beta^2 m}}{\beta^{n-k+1}}. \tag{5.115}$$

with

$$\boldsymbol{H}^*(\boldsymbol{x}) = \left(\frac{\partial^2 f(\boldsymbol{x})}{\partial x_i \partial x_j} - \frac{|\nabla f(\boldsymbol{x})|}{|\nabla g(\boldsymbol{x})|} \frac{\partial^2 g(\boldsymbol{x})}{\partial x_i \partial x_j} \right)_{i,j=k+1,\dots,n-1}. \tag{5.116}$$

Integrating over the manifold M yields the result of the theorem. □

An analogous result can be found for surface integrals.

Theorem 50 *Let condition A be fulfilled. Further is given a twice continuously differentiable function $f : \mathbb{R}^n \to \mathbb{R}$ and a continuous function $h : \mathbb{R}^n \to \mathbb{R}$. Assume further that:*

1. *The function f attains its global maximum $m = \max_{\boldsymbol{x} \in G} f(\boldsymbol{x})$ with respect to G exactly on the compact k-dimensional submanifold $M \subset G$.*
2. *In all points $\boldsymbol{x} \in M$ the $(n-k-1) \times (n-k-1)$-matrix is $\boldsymbol{H}^*(\boldsymbol{x})$ regular. This matrix is defined by*

$$\boldsymbol{H}^*(\boldsymbol{x}) = \boldsymbol{A}^T(\boldsymbol{x}) \boldsymbol{H}(\boldsymbol{x}) \boldsymbol{A}(\boldsymbol{x})$$

with

$$\boldsymbol{H}(\boldsymbol{x}) = \left(f^{ij}(\boldsymbol{x}) - \frac{|\nabla f(\boldsymbol{x})|}{|\nabla g(\boldsymbol{x})|} g^{ij}(\boldsymbol{x}) \right)_{i,j=1,\dots,n}$$

and

$$\boldsymbol{A}(\boldsymbol{x}) = (\boldsymbol{a}_1(\boldsymbol{x}), \dots, \boldsymbol{a}_{n-k-1}(\boldsymbol{x})).$$

Here the $\boldsymbol{a}_1(\boldsymbol{x}), \dots, \boldsymbol{a}_{n-k-1}(\boldsymbol{x})$ are an orthonormal basis of the subspace of $T_G(\boldsymbol{x}^*)$, which is orthogonal to $T_M(\boldsymbol{x}^*)$.

Then we have as $\beta \to \infty$ that

$$\int_G h(\boldsymbol{x}) e^{\beta^2 f(\boldsymbol{x})} ds_G(\boldsymbol{x}) \sim (2\pi)^{(n-k-1)/2} \int_M \frac{h(\boldsymbol{x})\, ds_M(\boldsymbol{x})}{\sqrt{|\det(\boldsymbol{H}^*(\boldsymbol{x}))|}} e^{\beta^2 m} \beta^{-(n-k-1)} \tag{5.117}$$

Here $ds_M(\boldsymbol{x})$ denotes surface integration over M.

PROOF: Analogous to the last theorem. Only the factor $|\nabla f(\boldsymbol{x})|^{-1} \beta^{-2}$ does not appear, since we integrate only over the surface. □

The next theorem is a modification of theorem 42, p. 59 for boundary maxima.

Theorem 51 *Let condition A be fulfilled. Further is given a twice continuously differentiable function $f : \mathbb{R}^n \to \mathbb{R}$, a continuously differentiable function $g_1 : \mathbb{R}^n \to \mathbb{R}$ and a continuous function $g_0 : \mathbb{R}^n \to \mathbb{R}$.*

Assume further that:

1. *The function f attains its global maximum with respect to F only at the point $\boldsymbol{x}^* \in G$.*

2. *$g_0(\boldsymbol{x}^*) \neq 0$ and $g_1(\boldsymbol{x}^*) = 0$.*

3. *At $\boldsymbol{x}^*$ the $(n-1) \times (n-1)$-matrix $\boldsymbol{H}^*(\boldsymbol{x})$ is regular. This matrix is defined by*

$$\boldsymbol{H}^*(\boldsymbol{x}^*) = \boldsymbol{A}^T(\boldsymbol{x}^*)\boldsymbol{H}(\boldsymbol{x}^*)\boldsymbol{A}(\boldsymbol{x}^*)$$

   with

$$\boldsymbol{H}(\boldsymbol{x}^*) = \left(f^{ij}(\boldsymbol{x}^*) - \frac{|\nabla f(\boldsymbol{x}^*)|}{|\nabla g(\boldsymbol{x}^*)|} g^{ij}(\boldsymbol{x}^*) \right)_{i,j=1,\dots,n}$$

   and

$$\boldsymbol{A}(\boldsymbol{x}^*) = (\boldsymbol{a}_1(\boldsymbol{x}^*), \dots, \boldsymbol{a}_{n-1}(\boldsymbol{x}^*)).$$

   Here the $\boldsymbol{a}_1(\boldsymbol{x}^*), \dots, \boldsymbol{a}_{n-1}(\boldsymbol{x}^*)$ form an orthonormal basis of $T_G(\boldsymbol{x}^*)$.

Then we have the following asymptotic relation

$$\int_F g_0(\boldsymbol{x})|g_1(\boldsymbol{x})| \exp(\beta^2 f(\boldsymbol{x}))\, d\boldsymbol{x} \tag{5.118}$$

$$\sim 2(2\pi)^{(n-2)/2} \frac{g_0(\boldsymbol{x}^*)}{|\nabla f(\boldsymbol{x}^*)|} \left| \frac{\boldsymbol{d}^T \boldsymbol{H}^-(\boldsymbol{x}^*)\boldsymbol{d}}{\det(\boldsymbol{H}^*(\boldsymbol{x}^*))} \right|^{1/2} e^{\beta^2 f(\boldsymbol{x}^*)} \beta^{-(n+2)},\ \beta \to \infty.$$

Here $\boldsymbol{d} = \nabla g_1(\boldsymbol{x}^*) - \langle \boldsymbol{n}(\boldsymbol{x}^*), \nabla g_1(\boldsymbol{x}^*) \rangle \boldsymbol{n}(\boldsymbol{x}^*)$ is the projection of $\nabla g_1(\boldsymbol{x}^*)$ onto $T_G(\boldsymbol{x}^*)$.

PROOF: The proof is similar to the proof of theorem 42, p. 59.

We give a short outline. We assume first that $\boldsymbol{x}^* = \mathbf{o}$, $f(\mathbf{o}) = 0$ and that the tangential space of G at $\mathbf{o}$ is spanned by the unit vectors $\boldsymbol{e}_1, \dots, \boldsymbol{e}_{n-1}$ and that the curvatures of the surface vanish at $\mathbf{o}$. The gradient $\nabla f(\mathbf{o})$ points in the direction of the unit vector $\boldsymbol{e}_n$.

We write then

$$I(\beta) = \int_F g_0(\boldsymbol{x})|g_1(\boldsymbol{x})| \exp(\beta^2 f(\boldsymbol{x}))\, d\boldsymbol{x} \tag{5.119}$$

For the function f we have then the following second order Taylor expansion at the origin in $\mathbb{R}^n$

$$f(\boldsymbol{x}) \sim \frac{1}{2} \sum_{i,j=1}^{n} f^{ij}(\mathrm{o}) x_i x_j + f^n(\mathrm{o}) x_n. \tag{5.120}$$

For the function g_1 we get from its first order Taylor expansion

$$g_1(\boldsymbol{x}) \sim \sum_{i=1}^{n} g_1^i(\mathrm{o}) x_i. \tag{5.121}$$

Replacing $g_0(\boldsymbol{x})$ by $g_0(\mathrm{o})$ we have approximately for the integral

$$I(\beta) \sim g_0(\mathrm{o}) \int_F |\sum_{i=1}^{n} g_1^i(\mathrm{o}) x_i| \exp\left(\beta^2 (\sum_{i,j=1}^{n} \frac{f^{ij}(\mathrm{o})}{2} x_i x_j + f^n(\mathrm{o}) x_n) \right) d\boldsymbol{x}. \tag{5.122}$$

Changing the integration domain to $\mathbb{R}^{n-1} \times [0, \infty)$ we have again approximately

$$I(\beta) \sim g_0(\mathrm{o}) \int_{\mathbb{R}^{n-1}} \left[\int_0^\infty |\sum_{i=1}^{n} g_1^i(\mathrm{o}) x_i| \right. \tag{5.123}$$

$$\left. \times \exp\left(\beta^2 (\frac{1}{2} \sum_{i,j=1}^{n} f^{ij}(\mathrm{o}) x_i x_j + f^n(\mathrm{o}) x_n) \right) dx_n \right] dx_1 \ldots dx_{n-1}.$$

Making the substitution $x_i \mapsto u_i = \beta x_i$ for $i = 1, \ldots, n-1$ and $x_n \mapsto u_n = \beta^2 x_n$ gives then

$$= \frac{g_0(\mathrm{o})}{\beta^{n+2}} \int_{\mathbb{R}^{n-1}} \left[\int_0^\infty |\sum_{i=1}^{n-1} g_1^i(\mathrm{o}) u_i + \beta^{-1} g_n(\mathrm{o}) u_n| \right. \tag{5.124}$$

$$\left. \times \exp\left(\sum_{i,j=1}^{n-1} \frac{f^{ij}(\mathrm{o})}{2} u_i u_j + \beta^{-1} \sum_{j=1}^{n} \frac{f^{nj}(\mathrm{o}) u_j u_n}{2} + f^n(\mathrm{o}) u_n \right) du_n \right] du_1 \ldots du_{n-1}.$$

As $\beta \to \infty$ we can neglect the terms multiplied by β^{-1} and we get then

$$I(\beta) \sim \frac{g_0(\mathrm{o})}{\beta^{n+2}} \int_{\mathbb{R}^{n-1}} \left[\int_0^\infty |\sum_{i=1}^{n-1} g_1^i(\mathrm{o}) u_i| \right. \tag{5.125}$$

$$\left. \times \exp\left(\frac{1}{2} \sum_{i,j=1}^{n-1} f^{ij}(\mathrm{o}) u_i u_j + f^n(\mathrm{o}) u_n \right) du_n \right] du_1 \ldots du_{n-1}$$

Splitting this into two integrals gives then

$$= \frac{g_0(o)}{\beta^{n+2}} \int\limits_{\mathbb{R}^{n-1}} |\sum_{i=1}^{n-1} g_1^i(o)u_i| \exp\left(\frac{1}{2}\sum_{i,j=1}^{n-1} f^{ij}(o)u_iu_j\right) du_1 \dots du_{n-1} \quad (5.126)$$

$$\times \int\limits_0^\infty \exp(f^n(o)u_n)\, du_n.$$

For the second integral we get as value $|\nabla f(o)|^{-1}$, since only the partial derivative in the direction of the x_n-axis is not equal to zero, which follows from the Lagrange multiplier rule.

The first integral can be calculated using lemma 26, p. 30 giving

$$\int\limits_{\mathbb{R}^{n-1}} |\sum_{i=1}^{n-1} g_1^i(o)u_i| \exp\left(\frac{1}{2}\sum_{i,j=1}^{n-1} f^{ij}(o)u_iu_j\right) du_1 \dots du_{n-1} \quad (5.127)$$

$$= 2(2\pi)^{(n-2)/2} \left|\frac{\sum_{i,j=1}^{n-1} g_1^i(o)g_1^j(o)\tilde{f}^{ij}}{\det(H^*(o))}\right|^{1/2}.$$

Here the matrix $(\tilde{f}^{ij})_{i,j=1,\dots,n-1}$ is the inverse of the matrix $H^*(o) = (f^{ij}(o))_{i,j=1,\dots,n-1}$.

To find the value of the first integral we therefore need this inverse. We assumed that this matrix is regular. We have then $H^*(o) = A^T(o)H(o)A(o)$ with $A(o) = (e_1, \dots, e_{n-1})$. Since for a regular matrix the generalized inverse is equal to the inverse we find

$$\begin{aligned} H^{*-1}(o) &= H^{*-}(o) \quad (5.128)\\ = (A^T(o)H(o)A(o))^- &= A(o)^-H(o)^-(A^T(o))^-. \end{aligned}$$

Since the column vectors of $A(o)$ are orthogonal, we have using lemma 5, p. 11 that $A^-(o) = A^T(o)$ and we get

$$A(o)^-H(o)^-(A^T(o))^- = A^T(o)H^-(o)A(o). \quad (5.129)$$

Therefore we get for the product in equation (5.127)

$$\sum_{i,j=1}^{n-1} g_1^i(o)g_1^j(o)\tilde{f}^{ij} = (\nabla g_1(o))^T A(o)^T H^-(o)A(o)\nabla g_1(o). \quad (5.130)$$

But since $A(o)\nabla g_1(o) = \nabla g_1(o) - \langle e_n, \nabla g_1(o)\rangle e_n$ (see lemma 2, p. 9), this is the projection of $\nabla g_1(o)$ onto the tangential space. This proves the theorem in this special case.

We get as result for the integral

$$I(\beta) \sim 2(2\pi)^{(n-2)/2} \frac{g_0(\mathbf{o})}{|\nabla f(\mathbf{o})|} \left| \frac{\mathbf{d}^T \mathbf{H}^-(\mathbf{o})\mathbf{d}}{\det(\mathbf{H}^*(\mathbf{o}))} \right|^{1/2} \beta^{-(n+2)}, \ \beta \to \infty. \tag{5.131}$$

The general case is shown by transforming into local coordinates. □

Theorem 52 *Let condition A be fulfilled. Further is given a twice continuously differentiable function $f : \mathbb{R}^n \to \mathbb{R}$, a continuously differentiable function $g_1 : \mathbb{R}^n \to \mathbb{R}$ and a continuous function $g_0 : \mathbb{R}^n \to \mathbb{R}$.*

Assume further that:

1. *The function f attains its global maximum with respect to G exactly at the point $\mathbf{x}^* \in G$.*
2. *$g_0(\mathbf{x}^*) \neq 0$ and $g_1(\mathbf{x}^*) = 0$.*
3. *At $\mathbf{x}^*$ the $(n-1) \times (n-1)$-matrix $\mathbf{H}^*(\mathbf{x})$ is regular. This matrix is defined*

$$\mathbf{H}^*(\mathbf{x}^*) = \mathbf{A}^T(\mathbf{x}^*)\mathbf{H}(\mathbf{x}^*)\mathbf{A}(\mathbf{x}^*)$$

with

$$\mathbf{H}(\mathbf{x}^*) = \left(f^{ij}(\mathbf{x}^*) - \frac{|\nabla f(\mathbf{x}^*)|}{|\nabla g(\mathbf{x}^*)|} g^{ij}(\mathbf{x}^*) \right)_{i,j=1,\dots,n}$$

and

$$\mathbf{A}(\mathbf{x}^*) = (\mathbf{a}_1(\mathbf{x}^*), \dots, \mathbf{a}_{n-1}(\mathbf{x}^*)).$$

Here the $\mathbf{a}_1(\mathbf{x}^*), \dots, \mathbf{a}_{n-1}(\mathbf{x}^*)$ are an orthonormal basis of $T_G(\mathbf{x}^*)$.

Then we have

$$\int_G g_0(\mathbf{x})|g_1(\mathbf{x})| \exp(\beta^2 f(\mathbf{x})) \, ds_G(\mathbf{x}) \tag{5.132}$$

$$\sim 2(2\pi)^{(n-2)/2} g_0(\mathbf{x}^*) \left| \frac{\mathbf{d}^T \mathbf{H}^-(\mathbf{x}^*)\mathbf{d}}{\det(\mathbf{H}^*(\mathbf{x}))} \right|^{1/2} e^{\beta^2 f(\mathbf{x}^*)} \beta^{-n}, \ \beta \to \infty.$$

Here $\mathbf{d} = \nabla g_1(\mathbf{x}^*) - \langle \mathbf{n}(\mathbf{x}^*), \nabla g_1(\mathbf{x}^*) \rangle \mathbf{n}(\mathbf{x}^*)$ is the projection of $\nabla g_1(\mathbf{x}^*)$ onto $T_G(\mathbf{x}^*)$.

PROOF: Analogous to the last theorem. Only the factor $|\nabla f(\mathbf{x}^*)|^{-1}\beta^{-2}$ does not appear, since we integrate only over the surface. □

The following theorems are modifications of theorem 44, p. 62.

Theorem 53 *Let condition A be fulfilled. Further is given a twice continuously differentiable function $f : \mathbb{R}^n \to \mathbb{R}$, a continuously differentiable function $f_1 : \mathbb{R}^n \to \mathbb{R}$ and a continuous function $h : \mathbb{R}^n \to \mathbb{R}$.*

Assume further that:

1. *The function f attains its global maximum with respect to F only at the point $\boldsymbol{x}^* \in G$.*

2. *At $\boldsymbol{x}^*$ the $(n-1)\times(n-1)$-matrix $\boldsymbol{H}^*(\boldsymbol{x}^*)$ is regular. This matrix is defined by:*
$$\boldsymbol{H}^*(\boldsymbol{x}^*) = \boldsymbol{A}^T(\boldsymbol{x}^*)\boldsymbol{H}(\boldsymbol{x}^*)\boldsymbol{A}(\boldsymbol{x}^*)$$
with
$$\boldsymbol{H}(\boldsymbol{x}^*) = \left(f^{ij}(\boldsymbol{x}^*) - \frac{|\nabla f(\boldsymbol{x}^*)|}{|\nabla g(\boldsymbol{x}^*)|} g^{ij}(\boldsymbol{x}^*)\right)_{i,j=1,\ldots,n}$$
and
$$\boldsymbol{A}(\boldsymbol{x}^*) = (\boldsymbol{a}_1(\boldsymbol{x}^*), \ldots, \boldsymbol{a}_{n-1}(\boldsymbol{x}^*)).$$
Here the $\boldsymbol{a}_1(\boldsymbol{x}^*), \ldots, \boldsymbol{a}_{n-1}(\boldsymbol{x}^*)$ form an orthonormal basis of the tangential space $T_G(\boldsymbol{x}^*)$.

Then the following asymptotic relation holds

$$\int_F h(\boldsymbol{x})\exp(\beta f_1(\boldsymbol{x}) + \beta^2 f(\boldsymbol{x}))\, d\boldsymbol{x} \quad (5.133)$$

$$\sim (2\pi)^{(n-1)/2} \frac{h(\boldsymbol{x}^*)\exp(-\frac{1}{2}\boldsymbol{c}^T\boldsymbol{H}^-(\boldsymbol{x}^*)\boldsymbol{c})}{|\nabla f(\boldsymbol{x}^*)|\sqrt{|\det(\boldsymbol{H}^*(\boldsymbol{x}^*))|}} e^{\beta f_1(\boldsymbol{x}^*)+\beta^2 f(\boldsymbol{x}^*)}\beta^{-(n+1)}, \ \beta \to \infty.$$

Here:

1. *$\boldsymbol{c} = \nabla f_1(\boldsymbol{x}^*) - \langle \boldsymbol{n}(\boldsymbol{x}^*), \nabla f_1(\boldsymbol{x}^*)\rangle \boldsymbol{n}(\boldsymbol{x}^*)$ is the projection of $\nabla f_1(\boldsymbol{x}^*)$ onto $T_G(\boldsymbol{x}^*)$.*

2. *$\boldsymbol{H}^-(\boldsymbol{x}^*)$ is the generalized inverse of $\boldsymbol{H}(\boldsymbol{x}^*)$.*

PROOF: The proof is analogous to the proof of theorem 44, p. 62. We give a short outline.

We assume that $\boldsymbol{x}^* = \boldsymbol{o}$, $f(\boldsymbol{o}) = f_1(\boldsymbol{o}) = 0$ and that the tangential space of G at $\boldsymbol{o}$ is spanned by the unit vectors $\boldsymbol{e}_1, \ldots, \boldsymbol{e}_{n-1}$. The gradient $\nabla f(\boldsymbol{o})$ points in the direction of the unit vector $\boldsymbol{e}_n$.

We write

$$I(\beta) = \int_F h(\boldsymbol{x})\exp(\beta f_1(\boldsymbol{x}) + \beta^2 f(\boldsymbol{x}))\, d\boldsymbol{x}. \quad (5.134)$$

For f and f_1 we have then the following second order Taylor expansion at o (for f_1 we need as in theorem 44 only the first derivatives)

$$\beta f_1(x) + \beta^2 f(x) \sim \beta \sum_{i=1}^{n} f_1^i(o) x_i \tag{5.135}$$

$$+\frac{\beta^2}{2}\left(\sum_{i,j=1}^{n-1} f^{ij}(o) x_i x_j + \sum_{k=1}^{n} f^{kn}(o) x_k x_n + 2 f^n(o) x_n\right).$$

Collecting the terms for each variable gives

$$\exp(\beta f_1(x) + \beta^2 f(x)) \tag{5.136}$$

$$\sim \exp\left(\beta \sum_{i=1}^{n-1} f_1^i(o) x_i + \beta^2 \left(\frac{1}{2} \sum_{i,j=1}^{n-1} f^{ij}(o) x_i x_j\right.\right.$$

$$\left.\left. + (f^n(o) + \frac{f_1^n(o)}{\beta}) x_n + \frac{1}{2} \sum_{k=1}^{n} f^{kn}(o) x_k x_n\right)\right).$$

We have then approximately for the integral with again replacing $h(x)$ by $h(o)$

$$I(\beta) \sim h(o) \int_F \exp\left(\beta \sum_{i=1}^{n-1} f_1^i(o) x_i + \frac{\beta^2}{2}\left(\sum_{i,j=1}^{n-1} f^{ij}(o) x_i x_j\right.\right. \tag{5.137}$$

$$\left.\left. +2(f^n(o) + \frac{f_1^n(o)}{\beta}) x_n) + \sum_{k=1}^{n} f^{kn}(o) x_k x_n\right) dx.$$

Making the substitution $x_i \to y_i = \beta x_i$ for $i = 1, \ldots, n-1$ and $x_n \to y_n = \beta^2 x_n$ and enlarging the integration domain to $\mathbb{R}^{n-1} \times [0, \infty)$ gives

$$I(\beta) \sim \frac{h(o)}{\beta^{n+1}} \int_{\mathbb{R}^{n-1}} \left[\int_0^{\infty} \exp\left(\sum_{i=1}^{n-1} f_1^i(o) y_i + \frac{1}{2} \sum_{i,j=1}^{n-1} f^{ij}(o) y_i y_j\right.\right. \tag{5.138}$$

$$\left.\left. +(f^n(o) + \frac{f_1^n(o)}{\beta}) y_n + \frac{1}{2\beta} \sum_{k=1}^{n} f^{kn}(o) y_k y_n\right) dy_n\right] dy_1 \ldots dy_{n-1}.$$

The terms multiplied by β^{-1} are asymptotically negligible giving then

$$I(\beta) \sim \frac{h(o)}{\beta^{n+1}}$$

$$\times \int_{\mathbb{R}^{n-1}} \left[\int_0^{\infty} \exp\left(\sum_{i=1}^{n-1} f_1^i(o) y_i + \frac{1}{2} \sum_{i,j=1}^{n-1} f^{ij}(o) y_i y_j) + f^n(o) y_n\right) dy_n\right] dy_1 \ldots dy_{n-1}$$

Writing this now as the product of two integrals

$$= \frac{h(o)}{\beta^{n+1}} \int_{\mathbb{R}^{n-1}} \exp\left[\sum_{i=1}^{n-1} f_1^i(o)y_i + \frac{1}{2}\sum_{i,j=1}^{n-1} f^{ij}(o)y_iy_j\right] dy_1 \dots dy_{n-1} \quad (5.139)$$

$$\times \int_0^\infty \exp(f^n(o)y_n)\, dy_n.$$

For the integral over y_n we get as result $|\nabla f(o)|^{-1}$, since $|\nabla f(o)| = |f^n(o)|$ and for the integral over $y_1, \dots, y_{n-1}$ we obtain using lemma 26, p. 30

$$\int_{\mathbb{R}^{n-1}} \exp\left[\sum_{i=1}^{n-1} f_1^i(o)y_i + \frac{1}{2}\sum_{i,j=1}^{n-1} f^{ij}(o)y_iy_j\right] dy_1 \dots dy_{n-1} \quad (5.140)$$

$$= \quad (2\pi)^{(n-1)/2} \exp(-\frac{1}{2}\boldsymbol{c}^T \boldsymbol{H}^-(o)\boldsymbol{c})|\det(\boldsymbol{H}^*(o))|^{-1/2}.$$

This gives finally

$$I(\beta) \sim (2\pi)^{(n-1)/2} h(o) \frac{\exp(-\frac{1}{2}\boldsymbol{c}^T \boldsymbol{H}^-(o)\boldsymbol{c})}{|\nabla f(o)|\sqrt{\det(\boldsymbol{H}^*(o))}} \beta^{-(n+1)}, \ \beta \to \infty. \quad (5.141)$$

This is the result in this special case. By making a coordinate transformation the general case can be brought in this form. If the curvatures at the maximum point do not vanish, we have to replace the second derivatives of the function f by the the corresponding second derivatives with respect to local coordinates (see lemma 14, p. 17). □

The corresponding result for surface integrals is given in the following theorem.

Theorem 54 *Let condition A be fulfilled. Further is given a twice continuously differentiable function $f : \mathbb{R}^n \to \mathbb{R}$, a continouosly differentiable function $f_1 : \mathbb{R}^n \to \mathbb{R}$ and a continuous function $h : \mathbb{R}^n \to \mathbb{R}$. Assume further that:*

1. *The function f attains its global maximum with respect to G only at the point $\boldsymbol{x}^* \in G$.*
2. *At $\boldsymbol{x}^*$ the $(n-1)\times(n-1)$-matrix $\boldsymbol{H}^*(\boldsymbol{x}^*)$ is regular. This matrix is defined by:*

$$\boldsymbol{H}^*(\boldsymbol{x}^*) = \boldsymbol{A}^T(\boldsymbol{x}^*)\boldsymbol{H}(\boldsymbol{x}^*)\boldsymbol{A}(\boldsymbol{x}^*)$$

with:

$$\boldsymbol{H}(\boldsymbol{x}^*) = \left(f^{ij}(\boldsymbol{x}^*) - \frac{|\nabla f(\boldsymbol{x}^*)|}{|\nabla g(\boldsymbol{x}^*)|} g^{ij}(\boldsymbol{x}^*)\right)_{i,j=1,\dots,n}$$

and

$$\boldsymbol{A}(\boldsymbol{x}^*) = (\boldsymbol{a}_1(\boldsymbol{x}^*), \dots, \boldsymbol{a}_{n-1}(\boldsymbol{x}^*)).$$

Here the $\boldsymbol{a}_1(\boldsymbol{x}^*), \dots, \boldsymbol{a}_{n-1}(\boldsymbol{x}^*)$ form an orthonormal basis of the tangential space $T_G(\boldsymbol{x}^*)$.

Then as $\beta \to \infty$ the following asymptotic relation holds

$$\int_G h(\boldsymbol{x}) \exp(\beta f_1(\boldsymbol{x}) + \beta^2 f(\boldsymbol{x}))d\boldsymbol{x} \tag{5.142}$$

$$\sim \quad (2\pi)^{(n-1)/2} h(\boldsymbol{x}^*) \frac{\exp(-\frac{1}{2}\mathbf{c}^T \boldsymbol{H}^-(\boldsymbol{x}^*)\mathbf{c})}{\sqrt{|\det(\boldsymbol{H}^*(\boldsymbol{x}^*))|}} e^{\beta f_1(\boldsymbol{x}^*)+\beta^2 f(\boldsymbol{x}^*)} \beta^{-(n-1)}.$$

Here

1. $\mathbf{c} = \nabla f_1(\boldsymbol{x}^*) - (\boldsymbol{n}(\boldsymbol{x}^*), \nabla f_1(\boldsymbol{x}^*))\boldsymbol{n}(\boldsymbol{x}^*)$ *is the projection of* $\nabla f_1(\boldsymbol{x}^*)$ *onto* $T_G(\boldsymbol{x}^*)$,
2. $\boldsymbol{H}^-(\boldsymbol{x}^*)$ *is the generalized inverse of* $\boldsymbol{H}(\boldsymbol{x}^*)$.

PROOF: Analogous to the last proof. □

Chapter 6

Approximations for Normal Integrals

6.1 Time-Invariant Reliability Problems

As outlined in the introduction in the time-invariant model random influences on a technical structure are modelled by an n-dimensional random vector $\boldsymbol{X} = (X_1, \ldots, X_n)$. The limit state function $g : \mathbb{R}^n \to \mathbb{R}$ describes the state of the system. If $g(\boldsymbol{x}) > 0$, the structure is intact, but if the random vector $\boldsymbol{X}$ has the realisation $\boldsymbol{X} = \boldsymbol{x}$. But if instead $g(\boldsymbol{x}) \leq 0$ the structure is defect for such a realization. Therefore we have two domains:

1. The safe domain $S = \{\boldsymbol{x}; g(\boldsymbol{x}) > 0\}$ and
2. the failure domain $F = \{\boldsymbol{x}; g(\boldsymbol{x}) \leq 0\}$.

The boundary of the failure domain $G = \{\boldsymbol{x}; g(\boldsymbol{x}) = 0\}$ is called the limit state surface. Since if the random vector $\boldsymbol{X}$ has a p.d.f. the probability content of the limit surface is zero, i.e. $\mathbb{P}(\boldsymbol{X} \in G) = 0$, in general it does not matter if the failure domain is defined as here by the set $\{\boldsymbol{x}; g(\boldsymbol{x}) \leq 0\}$ or $\{\boldsymbol{x}; g(\boldsymbol{x}) < 0\}$.

For systems with components the reliability of the system depends on the the reliability of the components. We consider a system S with n components $K_1, \ldots, K_n$. The simplest models here are parallel and series systems.

A parallel system with n components fails if all components $K_1, \ldots, K_n$ fail. The probability of failure is then

$$\mathbb{P}(S \text{ defect}) = \mathbb{P}(K_1 \text{ defect}, \ldots, K_n \text{ defect}). \tag{6.1}$$

If the components are independent, this gives

$$\mathbb{P}(S \text{ defect}) = \prod_{i=1}^{n} \mathbb{P}(K_i \text{ defect}). \tag{6.2}$$

Contrary to this a series system fails if at least one component fails. This gives

$$\mathbb{P}(S \text{ defect}) = \mathbb{P}(\text{at least one component } K_i \text{ defect}). \tag{6.3}$$

If the components are independent, we get

$$\mathbb{P}(S \text{ defect}) = 1 - \prod_{i=1}^{n} \mathbb{P}(K_i \text{ not defect}). \tag{6.4}$$

Starting from these simple systems more complex systems can be constructed. If the failure of the component K_i is described by a limit state function g_i of the random vector $\boldsymbol{X}$, the failure probability of a parallel system can be written as

$$\mathbb{P}(\cap_{j=1}^{m}\{g_j(\boldsymbol{X}) \leq 0\}) \tag{6.5}$$

and for a series system with the same components we get

$$\mathbb{P}(\cup_{j=1}^{m}\{g_j(\boldsymbol{X}) \leq 0\}). \tag{6.6}$$

Further results and applications can be found for example in [49], [73], [51] and [71]. For many complex systems it is difficult to compute the system reliability as function of the component reliabilities, but all systems can be reduced to a series systems of parallel subsystems or parallel system of series subsystems.

6.2 Linear Approximations (FORM Concepts)

6.2.1 The Hasofer/Lind reliability index

First Freudenthal [65] proposed to use the distance of F to the mean of the distribution as a measure for the reliability and to linearize limit state functions in the point, where the p.d.f. at the boundary of the failure domain is maximal. Further proposals for a reliability index gave Cornell [42] and Rosenblueth/Esteva [121]. The problem with both definitions was that it was necessary to calculate the moments of non-linear functions of $\boldsymbol{X}$ for determining the value of the index for a failure domain. It was therefore proposed to linearize the limit state function. Ditlevsen [47] pointed out that then the result may depend on the specific form of the limit state function and not only on the shape of F.

Later, in 1974 Hasofer and Lind [69] proposed to define a reliability index for failure domains, which is invariant with respect to different choices of the limit state function for a given failure domain. They considered a random vector $\boldsymbol{X} = (X_1, \ldots, X_n)$, which was centered and standardized, i.e.

$$\begin{aligned} \mathbb{E}(X_i) &= 0 \text{ for } i = 1, \ldots, n, & (6.7) \\ \operatorname{cov}(X_i, X_j) &= \delta_{ij} \text{ for } i = 1, \ldots, n. & (6.8) \end{aligned}$$

For an arbitrary domain $F \subset \mathbb{R}^n$ Hasofer and Lind defined as reliability index $\beta(F)$ of this domain

$$\beta(F) = \min_{\boldsymbol{x} \in F} |\boldsymbol{x}|. \tag{6.9}$$

This idea is formulated in the original article for arbitrary random vectors. But only in the case of standard normal random vectors there is a simple relation between this index and the probability content of half spaces. The p.d.f. of the standard normal distribution in $\mathbb{R}^n$ is

$$f(\boldsymbol{x}) = (2\pi)^{-n/2} \exp\left(-\frac{1}{2}\sum_{i=1}^{n} x_i^2\right) = (2\pi)^{-n/2} \exp\left(-\frac{1}{2}|\boldsymbol{x}|^2\right). \tag{6.10}$$

The level curves of this function are circles around the origin. If we now consider a domain F in $\mathbb{R}^n$, we have

$$\max_{\boldsymbol{x} \in F} f(\boldsymbol{x}) = f(\min_{x \in F} |\boldsymbol{x}|). \tag{6.11}$$

The maximum of the p.d.f. in F is at the points of F with minimal distance to the origin, i.e. whose euclidean norm is minimal.

In the case of a linear function

$$g_\beta(\boldsymbol{x}) = \beta - \boldsymbol{\alpha}^T \cdot \boldsymbol{x} \tag{6.12}$$

with $\sum_{i=1}^n \alpha_i^2 = 1$ and $\beta > 0$ due to the rotational symmetry of the standard normal distribution

$$\mathbb{P}(g_\beta(\boldsymbol{X}) \le 0) = \mathbb{P}(\beta \le \boldsymbol{\alpha}^T \cdot \boldsymbol{X}) = \mathbb{P}(\beta \le X_1) = \Phi(-\beta). \tag{6.13}$$

If we define $F_\beta = \{\boldsymbol{x}; g_\beta(\boldsymbol{x}) \le 0\}$, we have for these domains, which are half-spaces in $\mathbb{R}^n$ an one-to-one mapping between the probability $\mathbb{P}(\boldsymbol{X} \in F_\beta)$ and the reliability index $\beta(F_\beta) = \beta$ of F_β.

There are also corresponding inequalities between the probability contents and reliability indices of two half-spaces F_1 and F_2. If $\beta(F_1) = \beta_1$ and $\beta(F_2) = \beta_2$, then

$$\beta_1 > \beta_2 \Longleftrightarrow \mathbb{P}(F_1) < \mathbb{P}(F_2). \tag{6.14}$$

Here a larger reliability index corresponds to a smaller probability content and vice versa. But in general this is true only for such half-spaces.

For linear limit state functions we have this one-to-one relation between the reliability index and the probability content of the failure domain. If $\beta(F)$ is known, we get immediately $\mathbb{P}(F) = \Phi(-\beta(F))$.

But in the case of domains defined by non-linear functions the situation is different. Here we cannot compute the value of $\mathbb{P}(F)$ from the reliability index. As outlined for example in [89], p. 57 we can have failure domains F_a and F_b such that for the reliability indices we have $\beta(F_b) < \beta(F_a)$, but for the probabilities $\mathbb{P}(F_a) > \mathbb{P}(F_b)$. Therefore it is problematic to use this index for estimating the reliability of structures.

In the years 1975-1984 several attempts were made to generalize these relations between the index and probability contents to nonlinear functions.

The basic idea in this connection was to replace the nonlinear function g in the point $\boldsymbol{x}^1 \in F$ with minimal distance to the origin by a linear function g_L defined by

$$g_L(\boldsymbol{x}) = (\nabla g(\boldsymbol{x}^1))^T \cdot (\boldsymbol{x} - \boldsymbol{x}^1) \tag{6.15}$$

i.e. by its first order Taylor expansion at the point $\boldsymbol{x}^1$. The failure domain defined by this function is a half-space F_L and we have $\mathbb{P}(F_L) = \Phi(-|\boldsymbol{x}|) = \Phi(-\beta(F))$.

The proposed approximation method was therefore:

1. Calculate on the limit surface $G = \{\boldsymbol{x}; g(\boldsymbol{x}) = 0\}$ the point $\boldsymbol{x}^1$ with minimal distance to the origin, i.e. the point with $|\boldsymbol{x}^1| = \min\limits_{\boldsymbol{x} \in F} |\boldsymbol{x}|$.
2. Replace then g by the first order Taylor expansion given in equation (6.15) g_L at $\boldsymbol{x}^1$.
3. Approximate the probability $\mathbb{P}(F)$ by $\mathbb{P}(F_L) = \Phi(-|\boldsymbol{x}^1|)$.

The first problem here is: What is to be done if there are several points $\boldsymbol{x}^1, \ldots, \boldsymbol{x}^k$ on the limit state surface $|\boldsymbol{x}^1| = \ldots = |\boldsymbol{x}^k| = \min_{\boldsymbol{x} \in F} |\boldsymbol{x}|$. It may even happen that for whole surface parts M of G we have $|\boldsymbol{y}| = \min_{\boldsymbol{x} \in F} |\boldsymbol{x}|$ for all $\boldsymbol{y} \in M$.

For example consider the n-dimensional sphere with the center in the origin with radius β $F = \{\boldsymbol{x}; \beta^2 - \sum_{i=1}^n x_i^2 \leq 0\}$. Here we have for all points $\boldsymbol{y}$ on this sphere $|\boldsymbol{y}| = \min_{\boldsymbol{x} \in F} |\boldsymbol{x}|$. For this case an analytic solution is available, but what to do if only a part of the limit state surface is part of this sphere?

To solve the problem of several minimum points it was proposed to make a Taylor expansion g_L^i at all these points as in equation (6.15) and then to calculate $1 - \mathbb{P}(\cap_{i=1}^k \{\boldsymbol{x}; g_L^i(\boldsymbol{x}) \geq 0\})$. But this can be done only numerically. For the case of surface parts having minimal distance no useful approximations were proposed.

The essential problem was the quality of the approximations. It is possible to find any number of examples where the method is good and also any number of counterexamples with insufficient results. But such reasoning with examples can not replace a sound general mathematical justification.

In some papers (for example [61]) quadratic approximations instead of linear were considered. But it was unclear how to choose these approximations and if so an improvement is obtained. Only by applying methods of asymptotic analysis as described in the last chapter, this question can be answered.

6.2.2 Generalization to non-normal random variables

The concept of linear approximation can also be used for non-normal random vectors. This is done by transforming them into normal vectors.

If an n-dimensional random vector $\boldsymbol{X} = (X_1, \ldots, X_n)$ consists of n independent random variables $X_1, \ldots, X_n$ with c.d.f.'s $F_1(x_1), \ldots, F_n(x_n)$ and with positive continuous p.d.f.'s $f_1(x_1), \ldots, f_n(x_n)$, it can be transformed into a standard normal random vector $\boldsymbol{U}$ with independent components.

Such a transformation is given by

$$T : \mathbb{R}^n \to \mathbb{R}^n, \boldsymbol{x} \mapsto \boldsymbol{u} = \left(\Phi^{-1}\left[F_1(x_1)\right], \ldots, \Phi^{-1}\left[F_n(x_n)\right]\right). \tag{6.16}$$

The transformation determinant $\det(J_T(\boldsymbol{x}))$ is given by

$$\det(J_T(\boldsymbol{x})) = \prod_{i=1}^{n} \frac{f_i(x_i)}{\varphi_1(u_i)}. \tag{6.17}$$

This gives with transformation theorem for multivariate densities for the density $\tilde{f}(\boldsymbol{u})$ of the random vector $\boldsymbol{U}$ the form

$$\begin{aligned} \tilde{f}(\boldsymbol{u}) &= \prod_{i=1}^{n} f_i(x_i) |\det(J_T(\boldsymbol{x}))|^{-1} = \prod_{i=1}^{n} f_i(x_i) \prod_{i=1}^{n} \frac{\varphi_1(u_i)}{f_i(x_i)} \\ &= \prod_{i=1}^{n} \varphi_1(u_i) = (2\pi)^{-n/2} \exp\left(-\frac{1}{2}|\boldsymbol{u}|^2\right). \end{aligned} \tag{6.18}$$

By such a transformation the original failure domain $F = \{\boldsymbol{x}; g(\boldsymbol{x}) \leq 0\}$ is mapped into the domain $\boldsymbol{T}(F) = \{\boldsymbol{u}; g(\boldsymbol{T}^{-1}\boldsymbol{u}) \leq 0\}$. For this domain we can use the approximations described above. At the point $\boldsymbol{u}^1$ with minimal distance to the origin with respect to $\boldsymbol{T}(F)$ the function $g(\boldsymbol{T}^{-1}\boldsymbol{u})$ is replaced by its first order Taylor expansion and then for the domain bounded by this linear function approximations are made.

With this a method was developed for approximating arbitrary failure domains in the case of a random vector with indepedent components with positive continuous p.d.f's. If the p.d.f. is zero in some parts of $\mathbb{R}^n$, it is still possible to apply this transformation, but these parts might be mapped into infinity. This can cause problems for numerical algorithms which search the minimal distance points.

A further generalization was found by Hohenbichler and Rackwitz [72] in 1981 using the Rosenblatt-transformation. This transformation, which was described first by Rosenblatt in [120], is a transformation $\boldsymbol{T} : \mathbb{R}^n \to \mathbb{R}^n$, which maps an arbitrary n-dimensional random vector $\boldsymbol{X}$ with positive p.d.f. into a standard normal random vector $\boldsymbol{U}$ with independent components.

This transformation is defined by

$$\begin{aligned} u_1 &= \Phi^{-1}\left[F_1(x_1)\right] \\ u_2 &= \Phi^{-1}\left[F_2(x_2|x_1)\right] \\ &\vdots \\ u_i &= \Phi^{-1}\left[F_i(x_i|x_1, \ldots, x_{i-1})\right] \\ &\vdots \\ u_n &= \Phi^{-1}\left[F_n(x_n|x_1, \ldots, x_{n-1})\right]. \end{aligned} \tag{6.19}$$

Here $F_i(x_i|x_1, \ldots, x_{i-1})$ is the conditional c.d.f. of X_i under the condition $X_1 = x_1, \ldots, X_{i-1} = x_{i-1}$.

The problem in this approach is the calculation of the conditional c.d.f.'s in the case of dependent components. In [72] it was proposed to find them by numerical differentiation, but since such methods are numerically not very stable, no example is known to the author where this was done. Further the results may depend on the ordering of the random variables as was pointed out by Dolinski [52].

These methods described in this section are known as FORM (First Order Reliability Methods), since here only a first order Taylor expansion of the limit state function is made.

In the next section asymptotic approximations for normal random vectors are derived and in the next chapter it it shown that such transformations to normality are not necesssary for the derivation of asymptotic approximations.

6.3 SORM Concepts

As already said, additional to the FORM approximations based on linear Taylor expansions of the limit state function at the maximum point of the normal density, SORM (Second Order Reliability Methods) were studied in the late 70's (see [61]), where a second order Taylor expansion of the limit state function is made. That means that we use also the Hessian $\boldsymbol{H}_g(\boldsymbol{x})$ of the limit state function to fit an approximation.

In this context it is necessary to distinguish between two questions:

1. The calculation of the c.d.f. of a quadratic form of a normally distributed n-dimensional random vector $\boldsymbol{X}$

$$g_q(\boldsymbol{X}) = a + \boldsymbol{b}^T\boldsymbol{X} + \frac{1}{2}\boldsymbol{X}^T\boldsymbol{C}\boldsymbol{X}. \tag{6.20}$$

   with $a \in I\!R$, $\boldsymbol{b}$ an n-dimensional vector and $\boldsymbol{C}$ an $n \times n$ matrix.

2. The derivation of approximating quadratic forms $g_q(\boldsymbol{X})$ for arbitrary limit state functions $g(\boldsymbol{X})$ of a normal random vector $\boldsymbol{X}$, which give an " optimal " approximation for the failure probability.

These two problems are different. In the first case a quadratic form of normal random variables is given, i.e. it is known that the limit state function is exactly such a quadratic function, even in the regions far off from the minimal distance point.

Possible improvement are given by Tvedt ([133],[134] and [135]), Wu/Wirsching [141] and Torng/Wu [132]. They calculate instead of the asymptotic approximation for the probability content of the paraboloid the exact probability content by higher order expansions or transform methods. This gives better results if in fact the limit state surface is exactly a second order surface, elsewhere the effect of this method on the results depends on the problem. Naess [104] gives bounds for the distribution of quadratic forms.

In the second case the problem is to find for a limit state function which is *not a quadratic form* of a normal random vector such a quadratic form, which gives in some sense a good approximation for the failure probability, i.e. we would like that $\Pr(g(\boldsymbol{X}) \leq 0) \approx \Pr(g_q(\boldsymbol{X}) \leq 0)$.

In [89] in chapter 4, p. 66-69, for example, the two different problems are merged and an approximation method, which should improve the asymptotic concept outlined in the following, is proposed.

6.4 Asymptotic SORM for Normal Integrals

If we consider failure domains in the space of i.i.d. standard normal random variables, failure occurs in general, if at least some of the random influences have extreme values, for example if there are very large loads or if the resistances are small. For realizations near the mean vector $\mathbf{o}$ the structure normally is intact. Therefore we can assume in general the failure domain has a large distance to the origin.

Let be given a failure domain $\tilde{F}$ with distance β_0 to the origin. The failure probability $I\!P(\tilde{F})$ is then

$$I\!P(\tilde{F}) = \frac{1}{(2\pi)^{n/2}} \int_{\tilde{F}} \exp\left(-\frac{1}{2}\sum_{i=1}^{n} x_i^2\right) d\boldsymbol{x}. \tag{6.21}$$

Making the substitution $\boldsymbol{x} \mapsto \boldsymbol{y} = \beta_0^{-1}\boldsymbol{x}$, this integral over the domain $\tilde{F}$ is transformed into an integral over the domain $\beta_0^{-1}\tilde{F}$ such that now

$$I\!P(\tilde{F}) = \frac{\beta_0^n}{(2\pi)^{n/2}} \int_{\beta_0^{-1}\tilde{F}} \exp\left(-\frac{\beta_0^2}{2}\sum_{i=1}^{n} x_i^2\right) d\boldsymbol{x} \tag{6.22}$$

This is an integral over a domain with distance one to the origin.

Studying now the asymptotic behavior of the integral

$$I(\beta) = \frac{1 \cdot \beta^n}{(2\pi)^{n/2}} \int_{\beta_0^{-1}\tilde{F}} \exp\left(-\frac{\beta^2}{2}\sum_{i=1}^{n} x_i^2\right) d\boldsymbol{x}, \tag{6.23}$$

we get asymptotic approximations for $I\!P(\tilde{F})$, since $I(\beta_0) = I\!P(\tilde{F})$.

The domain $\beta_0^{-1}\tilde{F}$ is now taken as fixed. So the problem of approximating an integral with large distance β_0 to the origin is transformed into the problem of approximating a Laplace integral over a fixed integration domain as a parameter approaches infinity.

This integral $I(\beta_0)$ is imbedded in a sequence of integrals depending on a parameter β. For these integrals asymptotic approximations are derived as $\beta \to \infty$. If the parameter value β_0 is large at the integral of interest, the value of the asymptotic approximation for this integral will be a good approximation for this integral in general.

In this section the results of the last chapter are applied to normal integrals. Given is an n-dimensional integral in the form

$$(2\pi)^{-n/2} \int_F \exp\left(-\frac{1}{2}\sum_{i=1}^n x_i^2\right) d\boldsymbol{x} = (2\pi)^{-n/2} \int_F \exp\left(-\frac{1}{2}|\boldsymbol{x}|^2\right) d\boldsymbol{x}. \tag{6.24}$$

with $F \subset \mathbb{R}^n$.

Now we can construct a sequence $I(\beta)$ of integrals by defining the domains $\beta F = \{\boldsymbol{x}; \beta^{-1}\boldsymbol{x} \in F\}$ and setting

$$I(\beta) = (2\pi)^{-n/2} \int_{\beta F} \exp\left(-\frac{1}{2}|\boldsymbol{x}|^2\right) d\boldsymbol{x}. \tag{6.25}$$

With the substitution $\boldsymbol{x} \mapsto \boldsymbol{y} = \beta^{-1}\boldsymbol{x}$ this gives

$$I(\beta) = \frac{\beta^n}{(2\pi)^{n/2}} \int_F \exp\left(-\frac{\beta^2}{2}|\boldsymbol{x}|^2\right) d\boldsymbol{x}. \tag{6.26}$$

These integrals are Laplace integrals. The integration domain does not change. Studying the behavior of these integrals as $\beta \to \infty$, we get information about the asymptotic probability content of βF as $\beta \to \infty$.

Essential for the asymptotic behavior of these integrals is the structure of the domain F in this subset of F where the function $|\boldsymbol{x}|$ attains its global minimum with respect to F.

Using concepts of asymptotic analysis it is possible to obtain approximations also for the cases where the minimum is at several points or on a whole surface part.

If there are several points $\boldsymbol{x}^1, \dots, \boldsymbol{x}^k$ where the minimum occurs, F is split up into k subsets $F_1, \dots, F_k$ such that in each such subset lies one point. Then for each domain an approximation is derived and then they are added.

To simplify the following derivations we consider with the exception of theorem 56, p. 93 always only the case of a single minimum point or surface part.

The first result is about domains in $\mathbb{R}^n$, which are bounded by hyperplanes. It was published 1964 by Ruben [122].

Theorem 55 *Given are n affin-linear functions $g_i(\boldsymbol{x}) = \boldsymbol{a}_i^T(\boldsymbol{x} - \boldsymbol{x}^*)$ where the $\boldsymbol{a}_i$'s and $\boldsymbol{x}^*$ are constant vectors with $\det(\boldsymbol{a}_1, \dots, \boldsymbol{a}_n) \neq 0$ and $\boldsymbol{x}^* = \sum_{i=1}^n \gamma_i \boldsymbol{a}_i$ with $\gamma_i > 0$ for $i = 1, \dots, n$. For the set F defined by*

$$F = \cap_{i=1}^n \{\boldsymbol{x}; g_i(\boldsymbol{x}) \leq 0\}, \tag{6.27}$$

the following asymptotic approximation is obtained for $\mathbb{P}(\beta F)$ (= $I(\beta)$ in (6.25))

$$\mathbb{P}(\beta F) \sim \frac{\varphi_n(\beta \boldsymbol{x}^*)}{|\det(\boldsymbol{a}_1, \dots, \boldsymbol{a}_n)| \prod_{i=1}^n \gamma_i} \beta^{-n}, \beta \to \infty. \tag{6.28}$$

PROOF: See [33], corollary 1, p. 95. □

The first result for domains bounded by non-linear functions was published by the author in 1984 (see [23]). It is given in the form as it was published, i.e. for the case of several minimal distance points.

Theorem 56 *Given is a twice continuously differentiable function $g : \mathbb{R}^n \to \mathbb{R}$ such that for all points $\boldsymbol{x}$ on the surface $G = \{\boldsymbol{x}; g(\boldsymbol{x}) = 0\}$ always $\nabla g(\boldsymbol{x}) \neq \mathbf{o}$. The set F is defined by*

$$F = \{\boldsymbol{x}; g(\boldsymbol{x}) \leq 0\} \tag{6.29}$$

and further

$$\min_{\boldsymbol{x} \in F} |\boldsymbol{x}| = 1. \tag{6.30}$$

Assume that the following conditions are fulfilled:

1. *The function $|\boldsymbol{x}|$ attains its global minimum one with respect to the hypersurface $G = \{\boldsymbol{x}; g(\boldsymbol{x}) = 0\}$ exactly at k points $\boldsymbol{x}^1, \ldots, \boldsymbol{x}^k$.*
2. *At all these points $\boldsymbol{x}^1, \ldots, \boldsymbol{x}^k$ all main curvatures of the surface G are less than one.*

Then we have for $\mathbb{P}(\beta F)$ the asymptotic approximation

$$\mathbb{P}(\beta F) \sim \Phi(-\beta) \left[\sum_{i=1}^{k} \left(\prod_{j=1}^{n-1} (1 - \kappa_{ij})^{-1/2} \right) \right], \ \beta \to \infty. \tag{6.31}$$

Here the κ_{ij}'s ($j = 1, \ldots, n-1$) are the main curvatures of the surface G at the point $\boldsymbol{x}^i$.

PROOF: Proofs are given in [23] and [33]. Using the compactification lemma we can assume that F is a compact set.

We define $f(\boldsymbol{x}) = -\frac{1}{2}\sum_{i=1}^{n} x_i^2$ and use theorem 46, p. 67. Since $\mathbf{o} \notin F$, the maxima of f with respect to F are on the boundary of F. These are the points with minimal distance to the origin.

We assume first that there is only one such point $\boldsymbol{x}^1$. By a suitable rotation we can always achieve that $\boldsymbol{x}^1 = (0, \ldots, 0, 1)$ and that the main curvature directions of the surface G at this point $\boldsymbol{x}^1$ are the directions of the axes $\boldsymbol{e}_1, \ldots, \boldsymbol{e}_{n-1}$.

We have then

$$\mathbb{P}(\beta F) = (2\pi)^{-n/2} \int_{\beta F} \exp\left(-\frac{1}{2} \sum_{i=1}^{n} y_i^2 \right) d\boldsymbol{y}. \tag{6.32}$$

Making the transformation $\boldsymbol{y} \to \boldsymbol{x} = \beta^{-1}\boldsymbol{y}$ we get

$$= (2\pi)^{-n/2}\beta^n \int_{F} \exp\left(-\frac{\beta^2}{2} \sum_{i=1}^{n} x_i^2 \right) d\boldsymbol{x}. \tag{6.33}$$

Using the fact that $|\nabla f(\boldsymbol{x}^1)| = |\boldsymbol{x}^1| = 1$, we get with theorem 46, p. 67 the following approximation

$$\begin{aligned} \mathbb{P}(\beta F) &= \beta^n \int\limits_F (2\pi)^{-n/2} \exp\left(-\frac{\beta^2}{2}\sum_{i=1}^n x_i^2\right) d\boldsymbol{x} \qquad (6.34) \\ &\sim \frac{\exp(-\beta^2/2)}{\beta\sqrt{2\pi}} \frac{1}{\sqrt{|\det(\boldsymbol{H}^*(\boldsymbol{x}^1))|}}. \end{aligned}$$

Here, since $f^{ij}(\boldsymbol{x}^1) = -\delta_{ij}$, we get for the modified Hessian

$$\begin{aligned} \boldsymbol{H}^*(\boldsymbol{x}^1) &= \left(f^{ij}(\boldsymbol{x}^1) - \frac{|\nabla f(\boldsymbol{x}^1)|}{|\nabla g(\boldsymbol{x}^1)|} g^{ij}(\boldsymbol{x}^1)\right)_{i,j=1,\ldots,n-1} \qquad (6.35) \\ &= \left(-\delta_{ij} - \frac{g^{ij}(\boldsymbol{x}^1)}{|\nabla g(\boldsymbol{x}^1)|}\right)_{i,j=1,\ldots,n-1}. \end{aligned}$$

Further we obtain for the determinant

$$\begin{aligned} |\det(\boldsymbol{H}^*(\boldsymbol{x}^1))| &= \left|\det\left((-\delta_{ij} - \frac{g^{ij}(\boldsymbol{x}^1)}{|\nabla g(\boldsymbol{x}^1)|})_{i,j=1,\ldots,n-1}\right)\right| \qquad (6.36) \\ &= \left|\det\left((\delta_{ij} + \frac{g^{ij}(\boldsymbol{x}^1)}{|\nabla g(\boldsymbol{x}^1)|})_{i,j=1,\ldots,n-1}\right)\right|. \end{aligned}$$

Since due to our assumptions the main curvature directions are parallel to the axes, the mixed derivatives vanish. For the second derivatives (see equation (2.27), p. 15) we get so

$$g^{jj}(\boldsymbol{x}^1)|\nabla g(\boldsymbol{x}^1)|^{-1} = -\kappa_{1j} \text{ for } j = 1, \ldots, n-1. \qquad (6.37)$$

In equation (2.27), p. 15 the relation between the derivatives and the curvatures is given. Here κ_{1j} is the main curvature in the direction of the x_j-axis at the point $\boldsymbol{x}^1$.

Finally we get

$$\mathbb{P}(\beta F) \sim \frac{e^{-\beta^2/2}}{\sqrt{2\pi}\beta} \frac{1}{\sqrt{\prod_{j=1}^{n-1}(1-\kappa_{1j})}}, \ \beta \to \infty. \qquad (6.38)$$

Using Mill's ratio (2.89), p. 28 gives further

$$\mathbb{P}(\beta F) \sim \frac{\Phi(-\beta)}{\sqrt{\prod_{j=1}^{n-1}(1-\kappa_{1j})}}, \ \beta \to \infty. \qquad (6.39)$$

This result is independent of the chosen coordinate system. This gives the result for the case of one minimum point.

In the case of several minimum points the asymptotic approximation is calculated for each point and then these contributions are added. □

In the last equation we get the main curvatures of the surface since they are related to the first and second derivatives (equation (2.27), p. 15). The main curvatures in the minimal points must be less or equal 1, since elsewhere it would not be a minimal point anymore.

This result demonstrates the relevance of the second derivatives of the function g at the minimal points. For an useful approximation, which is exact in an asymptotic sense, i.e. the relative error approaches zero as $\beta \to \infty$, it is necessary to take in account the second derivatives.

In the sense of approximating the given function g by a simpler function this means that g is replaced by a quadratic function g_q in the form

$$g_q(\boldsymbol{x}) = (\beta - x_n) - \sum_{i=1}^{n-1} \frac{\kappa_i}{2} x_i^2. \tag{6.40}$$

(For details see [20] and [23]). This function defines an hyperparaboloid by $G_q = \{\boldsymbol{x}; g_q(\boldsymbol{x}) = 0\}$.

This gives the following method for the computation of asymptotic approximations for normal integrals in the case of a finite number of minimal points:

1. Calculate on the limit surface G all points $\boldsymbol{x}^1, \ldots, \boldsymbol{x}^k$ with $|\boldsymbol{x}^1| = \ldots = |\boldsymbol{x}^k| = \min_{\boldsymbol{x} \in G} |\boldsymbol{x}| = \beta_0$.

2. Calculate at all these points $\boldsymbol{x}^1, \ldots, \boldsymbol{x}^k$ the main curvatures κ_{ij} $(j = 1, \ldots, n-1)$ of the surface G.

3. If all $\kappa_{ij} < \beta_0^{-1}$, approximate $\mathbb{P}(F)$ by

$$\mathbb{P}(F) \approx \Phi(-\beta_0) \left[\sum_{i=1}^{k} \left(\prod_{j=1}^{n-1} (1 - \beta_0 \kappa_{ij})^{-1/2} \right) \right]. \tag{6.41}$$

The name SORM is used for all approximation methods wich use second order Taylor expansions of the limit state function. Now mainly this form of second order approximations based on the asymptotic considerations above is used.

The following theorem considers the case that the failure domain is the intersection of several domains defined by $g_1, \ldots, g_m$ and the minimum point lies on the intersection of some of the boundaries. This question is of interest in system reliability.

Theorem 57 *Let be given m twice continuously differentiable functions $g_1, \ldots, g_m : \mathbb{R}^n \to \mathbb{R}$ such that $g_i(o) > 0$ for at least one $i \in \{1, \ldots, m\}$. Let be defined $F = \cap_{i=1}^m \{y; g_i(y) \le 0\}$ and assume that there is only one point $y^* \in F$ where the function $|y|$ achieves its global minimum with respect to F. Assume that the following conditions are satisfied:*

1. *The gradients $a_i = \nabla g_i(y^*)$ are linearly independent.*
2. *For $i = 1, \ldots, k$ with $1 \le k \le m$ we have $g_i(y^*) = 0$ and for $j = k+1, \ldots, m$ instead $g_j(y^*) < 0$.*
3. *The vector y^* has a unique representation in the form*

$$y^* = -\sum_{i=1}^{k} \gamma_i a_i \tag{6.42}$$

   with $\gamma_i > 0$ for $i = 1, \ldots, k$.

4. *The $(n-k) \times (n-k)$-Matrix $H^*(y^*)$ is regular. Here the matrix $H^*(y^*)$ is defined by*

$$H^*(y^*) = A^T(y^*)H(y^*)A(y^*)$$

   with

   (a) $H(y^*) = \left(-\delta_{ij} - \sum_{s=1}^{k} \gamma_s g_s^{ij}(y^*)\right)_{i,j=1,\ldots,n}$,

   (b) *$A(y^*) = (a_{k+1}(y^*), \ldots, a_n(y^*))$. Here the $a_{k+1}(y^*), \ldots, a_n(y^*)$ are an orthonormal basis of the tangential space of the $(n-k)$-dimensional manifold $N = \cap_{i=1}^k \{y; g_i(y) = 0\}$ at y^*.*

5. *Let be defined $h_i(x) = (\nabla g_i(y^*))^T(y^* - x)$ for $i = 1, \ldots, k$. These functions define a domain $\tilde{F} = \cap_{i=1}^k \{x; h_i(x) \le 0\}$.*

Then the following asymptotic relation holds

$$\mathbb{P}(\beta F) \sim |\det(H^*(y^*))|^{-1/2} \mathbb{P}(\beta \tilde{F}),\ \beta \to \infty. \tag{6.43}$$

The probability $\mathbb{P}(\beta \tilde{F})$ in this equation can be approximated using theorem 55, p. 92.

PROOF: See corollary 3 in [33]. □

The meaning of this result is that, if the functions $g_1, \ldots, g_k$ are replaced by their linearizations in the minimal distance point y^*, one obtains an asymptotic approximation for the probability $\mathbb{P}(\beta F)$ is found by first approximating the linearized domain and then multiplying it by a factor determined by the curvatures of the manifold $\cap_{i=1}^k \{x; g_i(x) = 0\}$. In the case of a single limit state function, $\Phi(-\beta)$ is obtained by linearizing the limit state surface at the minimal distance point and then the factor $\prod_{j=1}^{n-1}(1 - \kappa_{1j})^{-1/2}$ gives the second order correction.

Theorem 58 *Given is a twice continuously differentiable function $g: \mathbb{R}^n \to \mathbb{R}$ with $g(\mathbf{o}) < 0$. Assume further that:*

1. *The set M of all points $\boldsymbol{x}$ in F with $|\boldsymbol{x}| = \min_{\boldsymbol{y} \in F} |\boldsymbol{y}| = 1$ is a k-dimensional submanifold of the $(n-1)$-dimensional manifold $G = \{\boldsymbol{x}; g(\boldsymbol{x}) = 0\}$.*

2. *For all $\boldsymbol{x} \in M$ we have $\nabla g(\boldsymbol{x}) \neq \mathbf{o}$.*

3. *For all $\boldsymbol{x} \in M$ the $(n-k-1) \times (n-k-1)$-matrix $\boldsymbol{H}^*(\boldsymbol{x})$ is regular. Here $\boldsymbol{H}^*(\boldsymbol{x}) = \boldsymbol{A}^T(\boldsymbol{x})\boldsymbol{H}(\boldsymbol{x})\boldsymbol{A}(\boldsymbol{x})$ with:*

   (a) $\boldsymbol{H}(\boldsymbol{x}) = \left(-\delta_{ij} - |\nabla g(\boldsymbol{x})|^{-1} g^{ij}(\boldsymbol{x})\right)_{i,j=1,\ldots,n}$.

   (b) $\boldsymbol{A}(\boldsymbol{x}) = (\boldsymbol{a}_{n-k+1}(\boldsymbol{x}), \ldots, \boldsymbol{a}_{n-1}(\boldsymbol{x}))$. Here the $\boldsymbol{a}_{n-k+1}(\boldsymbol{x}), \ldots, \boldsymbol{a}_{n-1}(\boldsymbol{x})$ are an orthonormal basis of the $(n-k-1)$-dimensional subspace of the $T_G(\boldsymbol{x})$ which is orthogonal to the k-dimensional tangential space $T_M(\boldsymbol{x})$.

Then the following asymptotic relation holds

$$\mathbb{P}(\beta F) \sim \tilde{Q}(\beta | k+1) \cdot \frac{\Gamma((k+1)/2)}{2\pi^{(k+1)/2}} \int_M \frac{ds_M(\boldsymbol{x})}{\sqrt{|\det(\boldsymbol{H}^*(\boldsymbol{x}))|}}, \quad \beta \to \infty. \tag{6.44}$$

Here $ds_M(\boldsymbol{x})$ denotes surface integration over M. $\tilde{Q}(x|k)$ is the complementary c.d.f. of the χ_k-distribution (see equation (2.91), p. 29).

PROOF: Using the compactification lemma we can assume that F is compact. We use theorem 49, p. 74. Here we have $f(\boldsymbol{x}) = -1/2 \sum_{i=1}^n x_i^2$ giving

$$\mathbb{P}(\beta F) = (2\pi)^{-n/2} \beta^n \int_F \exp(\beta^2 f(\boldsymbol{x}))\, d\boldsymbol{x}. \tag{6.45}$$

Now using the theorem we get, since $|\nabla f(\boldsymbol{x})| = 1$ for $\boldsymbol{x} \in M$ that

$$\mathbb{P}(\beta F) \sim e^{-\beta^2/2} \beta^{k-1} (2\pi)^{-(k+1)/2} \int_M \frac{ds_M(\boldsymbol{x})}{\sqrt{|\det(\boldsymbol{H}^*(\boldsymbol{x}))|}}. \tag{6.46}$$

With equation (2.91), p. 29 we get

$$\mathbb{P}(\beta F) \sim \tilde{Q}(\beta | k+1) \frac{\Gamma((k+1)/2)}{2\pi^{(k+1)/2}} \int_M \frac{ds_M(\boldsymbol{x})}{\sqrt{|\det(\boldsymbol{H}^*(\boldsymbol{x}))|}}. \tag{6.47}$$

This is the result of the theorem. □

The last result shows the influence of the dimension k of the set $M = \{y; y \in G, |y| = 1\}$ of minimal distance points on the asymptotic magnitude of $\mathbb{P}(\beta F)$. If $k = 0$, we have that this probability is of order $\Phi(-\beta)$, for $0 < k \leq n-1$ instead it is of order $\exp(-\beta^2/2)\beta^{k-1} \sim \Phi(-\beta)\beta^k$. Already in the papers Birndt/Richter [14] and Richter [117] the asymptotic order of the probability was derived, but not the exact value of the constant.

6.4.1 The generalized reliability index

1979 Ditlevsen [48] defined for failure domains F a generalized reliability index $\beta_G(F)$ by

$$\beta_G(F) = -\Phi^{-1}(\mathbb{P}(F)). \tag{6.48}$$

This index is equal to the number x on the real axis, where the standard normal integral from $-\infty$ to x has the value $1 - \mathbb{P}(F)$.

Apparently this it, contrary to the Hasofer/Lind index, an one-to-one mapping between $\mathbb{P}(F)$ and $\beta_G(F)$. It is strictly monotone

$$\beta_G(F_1) > \beta_G(F_2) \Leftrightarrow \mathbb{P}(F_1) < \mathbb{P}(F_2). \tag{6.49}$$

The definition of the generalized reliability index gives

$$\mathbb{P}(\beta F) = \Phi(-\beta_G(\beta F)). \tag{6.50}$$

For large values of $\beta_G(\beta F)$ Mill's ratio (2.89), p. 28 gives

$$\Phi(-\beta_G(\beta F)) \sim \frac{\exp(-\frac{1}{2}\beta_G(\beta F)^2)}{\sqrt{2\pi}\beta_G(\beta F))}. \tag{6.51}$$

This yields then

$$\ln(\mathbb{P}(\beta F)) \sim \frac{\ln(\exp(-\frac{1}{2}\beta_G(\beta F)^2)}{\sqrt{2\pi}\beta_G(\beta F)}. \tag{6.52}$$

Neglecting terms of lower order we get

$$\begin{aligned} \ln(\mathbb{P}(\beta F)) &\sim -\beta_G(\beta F)^2/2 \\ \beta_G(\beta F) &\sim \sqrt{2|\ln(\mathbb{P}(\beta F))|}. \end{aligned} \tag{6.53}$$

Here we see that the generalized reliability index is asymptotically proportional to the square root of $2\ln(\mathbb{P}(\beta F))$.

A problem is the use of this index in the engineering literature. In many cases, for example in optimization, we need the probability of failure and not the reliability index. But as outlined in chapter 3, example 3.2, p. 38 an asymptotic approximation for logarithm of a function does in general not give an asymptotic approximation the function if it is simply put in the exponent.

6.5 The Asymptotic Distribution in the Failure Domain

For further refining asymptotic approximations by importance sampling methods, it is useful to derive the asymptotic distribution of the random vector $\boldsymbol{X}$ in the failure domain, i.e. under the condition $g(\boldsymbol{X}) \leq 0$.

Let be defined by a twice continuously differentiable function g as usual a failure domain F and a limit state surface G. We consider the conditional distribution of $\boldsymbol{X}$ under the condition $g(\beta^{-1}\boldsymbol{X}) \leq 0$ for $\beta \to \infty$. Assume that:

1. The surface G has exactly one point $\boldsymbol{x}^*$ with $|\boldsymbol{x}^*| = \min_{\boldsymbol{x} \in G} |\boldsymbol{x}| = 1$,
2. At $\boldsymbol{x}^*$ we have for the gradient $\nabla g(\boldsymbol{x}^*) \neq \mathbf{o}$,
3. At $\boldsymbol{x}^*$ all main curvatures $\kappa_1, \ldots, \kappa_{n-1}$ of the surface G are less than unity.

By a suitable rotation of the coordinates we always can achieve that $\boldsymbol{x}^* = (0, \ldots, 0, 1)$ and the main curvature directions of the surface G at $\boldsymbol{x}^*$ are in the directions of the coordinate axes $\boldsymbol{e}_1, \ldots, \boldsymbol{e}_{n-1}$. Then the second mixed derivatives of g at $\boldsymbol{x}^*$ vanish. In a sufficiently small compact neighborhood $V \subset F$ of $\boldsymbol{x}^*$ we can introduce a local coordinate system $(h_1, \ldots, h_{n-1}, g)$ with the h_i's the arclengths in the directions of the main curvatures on the surface G at $\boldsymbol{x}^*$. Given is so an one-to-one mapping

$$T : V \to U, \boldsymbol{x} \mapsto \boldsymbol{u} = (h_1, \ldots, h_{n-1}, g) \tag{6.54}$$

with $U \subset \mathbb{R}^n$ being a neighborhood of the origin in $\mathbb{R}^n$.

We define a sequence of random vectors $\boldsymbol{X}_\beta$ by

$$\boldsymbol{X}_\beta = \frac{\boldsymbol{X} 1_{\beta V}}{\mathbb{P}(\boldsymbol{X} \in \beta V)}, \tag{6.55}$$

(1_A denotes the characteristic function of the set A).

From theorem 56, p. 93 we get that

$$\mathbb{P}(\boldsymbol{X} \in \beta F) = \mathbb{P}(g(\beta^{-1}\boldsymbol{X}) \leq 0) \sim \mathbb{P}(\boldsymbol{X} \in \beta V) \tag{6.56}$$

as $\beta \to \infty$. Therefore the asymptotic probability content of βF is equivalent to the content of βV as $\beta \to \infty$.

A further sequence $\boldsymbol{Y}_\beta$ of random vectors is defined by

$$\boldsymbol{Y}_\beta = (\beta h_1(\beta^{-1}\boldsymbol{X}_\beta), \ldots, \beta h_{n-1}(\beta^{-1}\boldsymbol{X}_\beta), -\beta^2 g(\beta^{-1}\boldsymbol{X}_\beta)). \tag{6.57}$$

This is a transformation of $\boldsymbol{X}_\beta$ into scaled local coordinates.

Theorem 59 *Under the conditions above we have for the the random vectors* $\boldsymbol{Y}_\beta$ *that*

$$\boldsymbol{Y}_\beta \xrightarrow{V} \boldsymbol{Y} \text{ as } \beta \to \infty. \tag{6.58}$$

Here $\boldsymbol{Y}$ *is an n-dimensional random vector with independent components. The distributions of these components are then*

$$Y_i \stackrel{V}{=} \mathrm{N}(0,(1-\kappa_i)^{-1}) \text{ for } i = 1,\dots,n-1, \tag{6.59}$$

$$Y_n \stackrel{V}{=} \mathrm{Exp}(|\nabla g(\boldsymbol{x}^*)|^{-1}). \tag{6.60}$$

PROOF: This theorem is proved using the Cramer-Wold-method (see theorem 29, p. 33). For this we consider the random variables $\exp(s\sum_{i=1}^n t_i Y_{\beta,i})$ with $\boldsymbol{t} = (t_1,\dots,t_n)$. The expected value of this variable is the value of the moment generating function of $\sum_{i=1}^n t_i Y_{\beta,i}$ at s. If we show that this function converges as $\beta \to \infty$ to the moment generating function of $\sum_{i=1}^n t_i Y_i$ for all t_i's for any s in an open interval with the origin in its interior, we have shown the convergence of the distribution of $\boldsymbol{Y}_\beta$ to $\boldsymbol{Y}$. The open interval may be different for different vectors $\boldsymbol{t}$.

We have for the moment generating function

$$\mathbb{E}\left[\exp(s\sum_{i=1}^n t_i Y_{\beta,i})\right] \tag{6.61}$$

$$= \frac{1}{(2\pi)^{n/2}\mathbb{P}(\boldsymbol{X}\in\beta V)}\int_{\beta V} \exp\left[\beta s\sum_{i=1}^{n-1} t_i h_i(\beta^{-1}\boldsymbol{y}) - \beta^2 s t_n g(\beta^{-1}\boldsymbol{y}) - \frac{1}{2}\sum_{i=1}^n y_i^2\right] d\boldsymbol{y}$$

$$= \frac{\beta^n}{(2\pi)^{n/2}\mathbb{P}(\boldsymbol{X}\in\beta V)}\int_V \exp\left[\beta\, \underbrace{s\sum_{i=1}^{n-1} t_i h_i(\boldsymbol{x})}_{=f_1} - \beta^2\, \underbrace{(s t_n g(\boldsymbol{x}) + \frac{1}{2}\sum_{i=1}^n x_i^2)}_{=f}\right] d\boldsymbol{x}.$$

From theorem 56, p. 93 we get as $\beta \to \infty$ that

$$\mathbb{P}(\boldsymbol{X}\in\beta V) \sim \frac{\Phi(-\beta)}{\sqrt{\prod_{i=1}^{n-1}(1-\kappa_i)}} \sim \frac{e^{-\beta^2/2}}{\sqrt{2\pi}\beta}\frac{1}{\sqrt{\prod_{i=1}^{n-1}(1-\kappa_i)}}. \tag{6.62}$$

For the derivatives of the functions f and f_1 at x^* we find

$$\frac{\partial f(x^*)}{\partial x_n} = st_n|\nabla g(x^*)| - 1, \tag{6.63}$$

$$\frac{\partial^2 f(x^*)}{\partial x_i \partial x_j} = \delta_{ij}(-1 - st_n g^{ij}(x^*)) \text{ for } i,j = 1,\dots,n-1, \tag{6.64}$$

$$\frac{\partial f_1(x^*)}{\partial x_i} = h_i'(x^*)st_i = st_i \text{ for } i = 1,\dots,n-1. \tag{6.65}$$

Since the h_i's are the arclengths along the main curvature directions at x^* we have $h_i'(x^*) = 1$.

Since f has a global maximum at x^* with respect to F, due to the Lagrange multiplier rule the first partial derivative with respect to x_n is negative and all other first derivatives are zero. From this follows if we take the interval around zero so small that $||\nabla g(x^*)|st_n| < 1$ we get then

$$|\nabla f(x^*)| = 1 - st_n|\nabla g(x^*)|. \tag{6.66}$$

For the matrix

$$\left(f^{ij}(x^*) - \frac{|\nabla f(x^*)|}{|\nabla g(x^*)|} g^{ij}(x^*)\right)_{i,j=1,\dots,n-1} \tag{6.67}$$

we get with these results for the elements

$$\begin{aligned} & f^{ij}(x^*) - \frac{|\nabla f(x^*)|}{|\nabla g(x^*)|} g^{ij}(x^*) \qquad (6.68)\\ = & -\delta_{ij}\left(1 + (st_n + \frac{1 - st_n|\nabla g(x^*)|}{|\nabla g(x^*)|})g^{ij}(x^*)\right) \\ = & -\delta_{ij}\left(1 + \frac{g^{ij}(x^*)}{|\nabla g(x^*)|}\right) = -\delta_{ij}(1-\kappa_i). \end{aligned}$$

Using theorem 53, p. 81 yields then

$$\begin{aligned} \mathbb{E}\left[\exp(s\sum_{i=1}^{n} t_i Y_{\beta,i})\right] & \sim \exp\left(s^2 \sum_{i=1}^{n-1}(1-\kappa_i)^{-1}\frac{t_i^2}{2}\right) \frac{t_n^{-1}|\nabla g(x^*)|^{-1}}{t_n^{-1}|\nabla g(x^*)|^{-1} - s} \qquad (6.69)\\ & = \exp\left(s^2 \sum_{i=1}^{n-1}(1-\kappa_i)^{-1}\frac{t_i^2}{2}\right) \frac{1}{1 - st_n^{-1}|\nabla g(x^*)|^{-1}}. \end{aligned}$$

This is the product of a moment generating function of a normal distribution with mean 0 and variance $\sum_{i=1}^{n-1} t_i^2(1-\kappa_i)^{-1}$ and an exponential distribution with mean $t_n|\nabla g(x^*)|$. The Cramer-Wold-device gives then the result. □

The meaning of this result becomes clearer if we consider some examples. If $\kappa_i = 0$ for $i = 1, \ldots, n-1$ the surface has no curvature at $\boldsymbol{x}^*$. The arc lengths are then simply the distances in the corresponding coordinate directions and they have a standard normal distribution. If $\kappa_i \to 1$ for $i = 1, \ldots, n-1$ the local shape of the surface approaches a spherical shape. The variances of the arc length $(1-\kappa_i)^{-1}$ go to infinity. If we consider the limit case of a sphere defined by the function $g_\beta(\boldsymbol{x}) = \beta - \sqrt{\sum_{i=1}^n x_i^2}$ we see that the conditional distribution of $\boldsymbol{X}$ under the condition $g_\beta(\boldsymbol{X}) = 0$ is the uniform distribution on the sphere. If instead $\kappa_i \to -\infty$ for $i = 1, \ldots, n-1$, the probability mass in the failure domain is concentrated more and more around $\boldsymbol{x}^*$ and the variances approach zero, since the surface is curved more from the origin into domains with smaller values of the p.d.f..

6.6 Numerical Procedures and Improvements

The main numerical problem in deriving such approximations is the computation of the minimal distance points. The first method developed by Rackwitz and Fiessler [110] specially for this problem is in fact a simplified modification of an algorithm of Pšeničnyj[109].

An overview of numerical algorithms for minimizing functions under constraints is given in the article of Liu/Der Kiureghian [87]. To avoid problems with numerical instabilities Der Kiureghian et al. [46] proposed to use instead of a paraboloid determined by the second derivatives at the minimum point one fitted through points in some distance to the minimal point. Further in [45] it is shown that the main curvatures of the limit state surface can be found from the final step lengths of numerical algorithms converging towards the minimal distance point.

The derived asymptotic approximations give in many cases sufficient approximations for small failure probabilities. But in some cases it may be necessary to improve the estimates. For this in the articles [125] and [75] was proposed to improve the estimate by importance sampling. Other articles about importance sampling methods are [36], [77] and [80]. Higher order approximations for normal integrals are derived in [35].

Further results in this direction, where the asymptotic form of the distribution in the failure domain as derived in the last section generalized to non-normal distributions as outlined in the next chapter, is used for importance Monte Carlo methods, can be found in [92] and [91].

6.7 Examples

The following two examples and figures are from the paper [23] in the Journal of the Engineering Mechanics Division, ASCE, Vol. 110 and are reprinted with permission.

EXAMPLE: In the two dimensional space is given a sequence of elliptical domains by the function $g_\beta(\boldsymbol{x}) = \beta^2 - ((x_1/a)^2 + x_2^2)$ with $a > 1$. On the curve $g_\beta(\boldsymbol{x}) = 0$ there are two points with minimal distance to the origin $\boldsymbol{x}_1 = (0, 1)$ and $\boldsymbol{x}_2 = (0, -1)$.

Using theorem 56, p. 93 gives the following approximation $Q(\beta)$ for $P(\beta) = \mathbb{P}(g_\beta(\boldsymbol{x}) \leq 0)$

$$Q(\beta) \sim \frac{2\Phi(-\beta)}{\sqrt{1 - a^{-2}}}. \tag{6.70}$$

In the picture the approximation $Q(\beta)$ is compared with the exact probability $P(\beta)$ for the values $a = 1.25$, 2 and 4. □

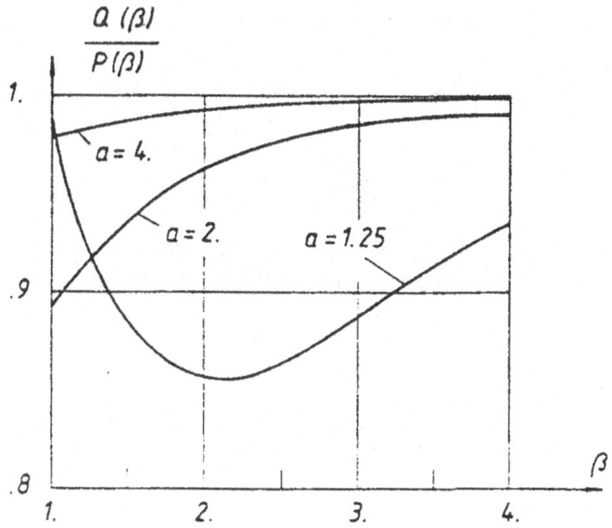

Figure 6.1: Probability content of an ellipse

EXAMPLE: Given are n i.i.d. exponential random variables $Y_1, \ldots, Y_n$ each having c.d.f. $F(y) = 1 - e^{-y}$. Further is given a limit state function

$$g(\boldsymbol{y}) = n + \alpha\sqrt{n} - \sum_{i=1}^{n} y_i. \tag{6.71}$$

The transformation into standard normally distributed variables is given by $x_i = -\Phi^{-1}\left[\exp(-y_i)\right]$ for $i = 1, \ldots, n$.

In the transformed space the function g has the form

$$\tilde{g}(\boldsymbol{x}) = \sum_{i=1}^{n} \ln[\Phi(-x_i)] + n + \alpha\sqrt{n}. \tag{6.72}$$

With the Lagrange multiplier method we find the only minimal point z_1 on the limit state surface $\tilde{G} = \{\boldsymbol{x}; \tilde{g}(\boldsymbol{x}) = 0\}$, which is

$$\boldsymbol{z}_1 = \left(-\Phi(e^{-\alpha/\sqrt{n}} - 1), \ldots, -\Phi(e^{-\alpha/\sqrt{n}} - 1)\right). \tag{6.73}$$

The distance of this point to the origin is $\sqrt{n}\Phi(e^{-\alpha/\sqrt{n}} - 1))$. The curvature of the surface at this point is in all directions constant and equal to $\frac{\varphi_1(z)}{\Phi(-z)} - z$ where $z = \Phi(e^{-\alpha/\sqrt{n}} - 1)$.

Then we get with theorem 56, p. 93 the approximation

$$\Phi(-\sqrt{n}z)\left(1 - z(\frac{\varphi_1(z)}{\Phi(-z)} - z)\right)^{-(n-1)/2}. \tag{6.74}$$

In the following picture this approximation is compared with the exact probability. for $0 \leq \alpha \leq 3$ and $n = 10$. and with the linear approximation $\Phi(-|\boldsymbol{z}_1|)$. □

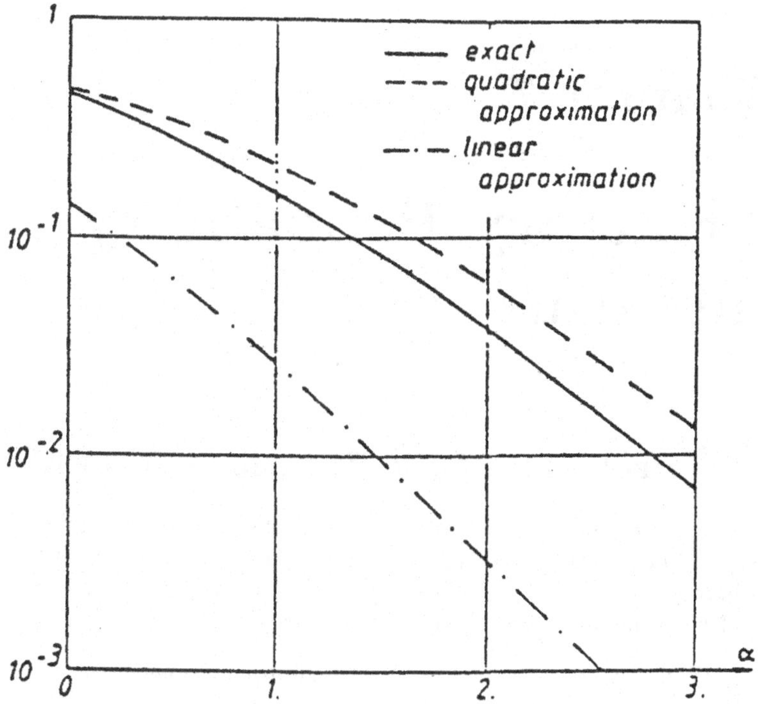

Figure 6.2: Sum of exponential random variables

Chapter 7

Arbitrary Probability Integrals

7.1 Problems of the Transformation Method

In the last chapter it was described how to transform non-normal random vectors into standard normal vectors to obtain asymptotic approximations for failure probabilities.

Two points of this method were unclear:

1. Why is it not possible to calculate approximations in the original space of random variables?

2. What is the meaning of the parameters of the solution in the standard normal space in relation the the original random variables?

Concerning the first point it was outlined by the author ([26] and [29]) that such methods can be used also in the original space of the random variables. This solves the second question too, then now the parameters of the solution depend on the distribution parameters of these random variables.

To understand the transformation method and to find this solution it is only necessary to clarify which function is minimized in this method.

An error in the transformation method appears to the author a misunderstanding of the mathematical meaning of the word " transformation". There are two different types of transformations, speaking loosely, in mathematics.

The first type changes the structure of the problem; for example to compute the product of two numbers a and b, by taking logarithms the problem of multiplication is transformed into a problem of addition, since, as commonly known $\ln(a \cdot b) = \ln(a) + \ln(b)$.

On the other hand the second type of transformation does not change the structure, but makes only a reparametrization. For example, if an integral over a domain in the two-dimensional space in cartesian coordinates is transformed into an integral in polar coordinates with the integral transformation theorem. Here only the coordinates are different. Maybe by such a reparametrization some functions appear in a simpler form, but this must correspond to some properties of the structure before the transformation.

The basic misconception of the transformation method is the belief that this is a transformation of the first type, but in reality it is of the second type, since the structure is not changed. Only a reparametrization is made, the problem of integration remains the same.

If we understand this, it becomes apparent that all results and methods of the transformation method *must have a natural interpretation in terms of the original random variables and their parameters.* To understand, what is done in the minimization procedure of the method, we have only to study the structure of the function, which is minimized. The correspondence between u_i and x_i is given by

$$u_i = \Phi^{-1}(F_i(x_i)). \tag{7.1}$$

But the c.d.f. $F_i(x_i)$ can be written as an integral over the exponent of the log likelihood function $l_i(x_i) = \ln(f_i(x_i))$ as

$$F_i(x_i) = \int_{-\infty}^{x_i} \exp(l_i(y))\, dy. \tag{7.2}$$

We get so for u_i that

$$u_i = \Phi^{-1}\left[\int_{-\infty}^{x_i} \exp(l_i(y))\, dy\right]. \tag{7.3}$$

With the relation $\Phi^{-1}(x) = -\Phi^{-1}(1-x)$, this can be rewritten as

$$u_i = -\Phi^{-1}\left[1 - \int_{-\infty}^{x_i} \exp(l_i(y))\, dy\right]. \tag{7.4}$$

Since $\int_{-\infty}^{\infty} \exp(l_i(y))dy = \int_{-\infty}^{\infty} f_i(x_i)dx_i = 1$, this yields

$$u_i = -\Phi^{-1}\left[\int_{x_i}^{\infty} \exp(l_i(y))\, dy\right]. \tag{7.5}$$

As $x_i \to \infty$ this integral can be approximated by the Laplace method (see corollary 37, p. 50), giving

$$\int_{x_i}^{\infty} \exp(l_i(y))dy \sim \int_{x_i}^{\infty} \exp(l_i(x_i) + l_i'(x_i)(y - x_i))dy = \frac{\exp(l_i(x_i))}{|l_i'(x_i)|}. \tag{7.6}$$

Inserting this into equation (7.5) we obtain

$$u_i \sim -\Phi^{-1}\left[\frac{\exp(l_i(x_i))}{|l_i'(x_i)|}\right]. \tag{7.7}$$

Or, in another form

$$\frac{\exp(l_i(x_i))}{|l_i'(x_i)|} \sim \Phi(-u_i). \tag{7.8}$$

Using Mill's ratio $\Phi(-y) \sim \varphi_1(y)/y$ as $y \to \infty$ yields then

$$\frac{\exp(l_i(x_i))}{|l_i'(x_i)|} \sim \frac{\varphi_1(u_i)}{u_i}. \tag{7.9}$$

By taking logarithms and neglecting terms of lower order

$$\exp(l_i(x_i)) \quad \sim \quad \exp\left(-\frac{u_i^2}{2}\right) \tag{7.10}$$

$$u_i^2 \quad \sim \quad -2 \cdot l_i(x_i). \tag{7.11}$$

If the u_i's are positive and tend towards infinity, we get for the norm of the vector $\boldsymbol{u}$ the relation

$$| \boldsymbol{u} |= \sqrt{\sum_{i=1}^{n} u_i^2} \sim \sqrt{-2\sum_{i=1}^{n} l_i(x_i)} = \sqrt{-2 \cdot l(\boldsymbol{x})}. \tag{7.12}$$

This can be stated as follows:

Minimizing the norm $| \boldsymbol{u} |= |T(\boldsymbol{x})|$ *in the standard normal space is asymptotically equivalent to maximizing the log likelihood function* $l(\boldsymbol{x})$ *in the original space.*

If we consider the log likelihood function in the standard normal space, this becomes clear

$$\begin{aligned} l(\boldsymbol{u}) &= -\frac{n}{2}\ln(2\pi) - \frac{1}{2}\sum_{i=1}^{n} u_i^2 \qquad (7.13) \\ &= -\frac{n}{2}\ln(2\pi) - \frac{1}{2} | \boldsymbol{u} |^2 . \end{aligned}$$

This shows that in fact the minimization of the distance to the origin is equivalent to the maximization of the log likelihood. Since through the transformation into the standard normal space only the form of the failure domain is altered, but not the basic structure of the problem, it should be clear now that in general the maximization of the log likelihood function in the transformed space is in principle the same as in the original space.

Knowing this it is possible to use asymptotic methods also in the original space. Here we do not have a geometric interpretation for the structure of the solution, but a probabilistic.

7.2 Asymptotic Approximations in the Original Space

7.2.1 The construction of Laplace integrals

The method described in the following is based on proposals of Freudenthal [65], Hasofer/Lind [69] and Shinozuka [126] for making analytic approximations of the limit state functions. Here with methods of asymptotic analysis a sound mathematical foundation is developed (see [26], [29] and [91]).

This method is especially important for random vectors with dependent components, since in this case the Rosenblatt-transformation is not applicable in practice.

Consider now a failure domain F in n-dimensional space where the log likelihood function of the distribution is given by $l(\boldsymbol{x})$. To obtain an asymptotic approximation for $\mathbb{P}(F)$, first all points in F are located, where the log likelihood function is maximal. Since F is a failure domain, these points will lie in general on the limit state surface G, the boundary of F.

For simplicity we assume that there is only one point $\boldsymbol{x}^1 \in G$ with $l(\boldsymbol{x}^1) = \max_{\boldsymbol{x}\in F} l(\boldsymbol{x})$ and that for all other points $\boldsymbol{x} \in F$ always $l(\boldsymbol{x}^1) > l(\boldsymbol{x})$.

If there are several points $\boldsymbol{x}^1, \ldots, \boldsymbol{x}^k$ on G, where the maximum is achieved, the failure domain is split up into k disjoint domains $F_1, \ldots, F_k$ with $\cup F_i = F$ such that in each F_i lies one $\boldsymbol{x}_i$. Then for each F_i with the following method an approximation for $\mathbb{P}(F_i)$ is computed and the sum of the approximations gives an approximation for $\mathbb{P}(F)$.

We define now

$$\beta_0 = \sqrt{-l(\boldsymbol{x}^1)} = \sqrt{-\max_{\boldsymbol{x}\in F} l(\boldsymbol{x})}. \tag{7.14}$$

Since F is a failure domain, where the p.d.f. is small in general, it can be assumed that $\max_{\boldsymbol{x}} l(\boldsymbol{x}) < 0$. Elsewhere it would be doubtful to use asymptotic methods anyway. Then the failure probability can be written as

$$\mathbb{P}(F) = \int_F \exp\left(\beta_0^2 \frac{l(\boldsymbol{x})}{\beta_0^2}\right) \, d\boldsymbol{x}. \tag{7.15}$$

We define now a scaled log likelihood function by

$$h(\boldsymbol{x}) = \frac{l(\boldsymbol{x})}{\beta_0^2}. \tag{7.16}$$

Then the integral $\mathbb{P}(F)$ can be written in the form

$$\mathbb{P}(F) = \int_F \exp(\beta_0^2 h(\boldsymbol{x})) \, d\boldsymbol{x}. \tag{7.17}$$

This is a Laplace type integral. If we consider the function $J(\beta)$, defined by

$$J(\beta) = \int_F \exp(\beta^2 h(\boldsymbol{x}))\, d\boldsymbol{x}, \tag{7.18}$$

for this function asymptotic approximations as $\beta \to \infty$ can be derived. The approximation value for β_0 gives an approximation for $\mathbb{P}(F)$.

Theorem 60 *Given is a p.d.f. $f : \mathbb{R}^n \to \mathbb{R}$ with $f(\boldsymbol{x}) > 0$ for all $\boldsymbol{x} \in \mathbb{R}^n$. The log likelihood function $\ln(f(\boldsymbol{x}))$ is denoted by $l(\boldsymbol{x})$. A twice continuously differentiable function $g : \mathbb{R}^n \to \mathbb{R}$ defines the failure domain $F = \{\boldsymbol{x}; g(\boldsymbol{x}) \leq 0\}$ with boundary $G = \{\boldsymbol{x}; g(\boldsymbol{x}) = 0\}$ and $\nabla g(\boldsymbol{x}) \neq o$ for all $\boldsymbol{x} \in G$.*

Assume further that:

1. *The function $l(\boldsymbol{x})$ is twice continuously differentiable.*
2. *The function $l(\boldsymbol{x})$ attains its maximum with respect to F only at the point $\boldsymbol{x}^* \in G$. The conditions about the derivatives of l and g in theorem 46, p. 67 are fulfilled.*
3. $\max_F l(\boldsymbol{x}) < o$.
4. *The conditions of the compactification lemma 38, p. 53 are fulfilled for the integrals*

$$\int_F \exp(\beta^2 h(\boldsymbol{x}))d\boldsymbol{x} \tag{7.19}$$

where $h(\boldsymbol{x}) = l(\boldsymbol{x})/\beta_0^2$ with $\beta_0 = \sqrt{-\max_{\boldsymbol{x}\in F} l(\boldsymbol{x})}$.

Then as $\beta \to \infty$ the following asymptotic relation is valid

$$\begin{aligned}\int_F \exp(\beta^2 h(\boldsymbol{x}))d\boldsymbol{x} &\sim (2\pi)^{(n-1)/2} \frac{e^{\beta^2 h(\boldsymbol{x}^*)}}{\sqrt{|(\nabla l(\boldsymbol{x}^*))^T \boldsymbol{C}(\boldsymbol{x}^*) \nabla l(\boldsymbol{x}^*)|}} \left(\frac{\beta_0}{\beta}\right)^{n+1} \\ &= (2\pi)^{(n-1)/2} \frac{e^{\beta^2 h(\boldsymbol{x}^*)}}{|\nabla l(\boldsymbol{x}^*)| \sqrt{|\det(\boldsymbol{A}^T(\boldsymbol{x}^*)\boldsymbol{H}(\boldsymbol{x}^*)\boldsymbol{A}(\boldsymbol{x}^*))|}} \left(\frac{\beta_0}{\beta}\right)^{n+1}\end{aligned} \tag{7.20}$$

Here $\boldsymbol{C}(\boldsymbol{x}^*)$ is the matrix of the cofactors of the matrix $\boldsymbol{H}(\boldsymbol{x}^*)$, i.e.

$$\boldsymbol{H}(\boldsymbol{x}^*) = \left(\frac{\partial^2 l(\boldsymbol{x}^*)}{\partial x_i \partial x_j} - \frac{|\nabla l(\boldsymbol{x}^*)|}{|\nabla g(\boldsymbol{x}^*)|} \frac{\partial^2 g(\boldsymbol{x}^*)}{\partial x_i \partial x_j} \right)_{i,j=1,\ldots,n}. \tag{7.21}$$

The matrix $\boldsymbol{A}(\boldsymbol{x}^*) = (a_1(\boldsymbol{x}^*), \ldots, a_{n-1}(\boldsymbol{x}^*))$ is an $n \times (n-1)$-matrix. The vectors $a_1(\boldsymbol{x}^*), \ldots, a_{n-1}(\boldsymbol{x}^*)$ are an orthonormal basis of the tangential space $T_G(\boldsymbol{x}^*)$.

PROOF: The result follows immediately from theorem 46, p. 67 if we set $f(\boldsymbol{x}) = h(\boldsymbol{x})$ and note that $h(\boldsymbol{x}) = \beta_0^{-2} l(\boldsymbol{x})$, see [26]. □

For the probability $\mathbb{P}(F)$ we get as approximation

$$\mathbb{P}(F) \sim \frac{(2\pi)^{(n-1)/2} f(\boldsymbol{x}^*)}{|\nabla l(\boldsymbol{x}^*)| \sqrt{|\det(\boldsymbol{H}^*(\boldsymbol{x}^*))|}}. \tag{7.22}$$

Here $\boldsymbol{H}^*(\boldsymbol{x}^*) = \boldsymbol{A}^T(\boldsymbol{x}^*)\boldsymbol{H}(\boldsymbol{x}^*)\boldsymbol{A}(\boldsymbol{x}^*)$.

In a similar way the result of theorem 57, p. 95 can be formulated for arbitrary probability densities.

Theorem 61 *Given are m twice continuously differentiable functions $g_i : \mathbb{R}^n \to \mathbb{R}$. These functions define the set $F = \cap_{i=1}^m \{\boldsymbol{x}; g_i(\boldsymbol{x}) \leq 0\}$.*

We assume further that:

1. *The function l attains its global maximum with respect to F only at the point $\boldsymbol{x}^* \in F$ and $\max_F l(\boldsymbol{x}) < 0$.*

2. *There is a $k \in \{1, \ldots, m\}$ with*

$$\begin{aligned} g_i(\boldsymbol{x}^*) &= 0 \text{ for } i = 1, \ldots, k \text{ and} \\ g_j(\boldsymbol{x}^*) &< 0 \text{ for } j = k+1, \ldots, m, \end{aligned} \tag{7.23}$$

and the gradients $\nabla g_i(\boldsymbol{x}^*)$ $(i = 1, \ldots, k)$ are linearly independent.

3. *The gradient $\nabla l(\boldsymbol{x}^*)$ has a unique representation in the form*

$$\nabla l(\boldsymbol{x}^*) = \sum_{i=1}^k \gamma_i \nabla g_i(\boldsymbol{x}^*) \tag{7.24}$$

with $\gamma_i > 0$ for $i = 1, \ldots, k$.

4. *The matrix $\boldsymbol{H}^*(\boldsymbol{x}^*)$ is regular. Here we have*

$$\boldsymbol{H}^*(\boldsymbol{x}^*) = \boldsymbol{A}^T(\boldsymbol{x}^*)\boldsymbol{H}(\boldsymbol{x}^*)\boldsymbol{A}(\boldsymbol{x}^*) \tag{7.25}$$

with

$$\boldsymbol{H}(\boldsymbol{x}^*) = \left(l^{ij}(\boldsymbol{x}^*) - \sum_{r=1}^k \gamma_r g_r^{ij}(\boldsymbol{x}^*) \right)_{i,j=1,\ldots,n} \tag{7.26}$$

and

$$\boldsymbol{A}(\boldsymbol{x}^*) = (\boldsymbol{a}_1(\boldsymbol{x}^*), \ldots, \boldsymbol{a}_{n-k}(\boldsymbol{x}^*)). \tag{7.27}$$

The $\boldsymbol{a}_1(\boldsymbol{x}^*), \ldots, \boldsymbol{a}_{n-k}(\boldsymbol{x}^*)$ form an orthonormal basis of the subspace $[\text{span}(g_1(\boldsymbol{x}^*), \ldots, \nabla g_k(\boldsymbol{x}^*))]^\perp$ in $\mathbb{R}^n$.

5. *The conditions of the compactification lemma 38, p. 53 are fulfilled for the integrals*

$$\int_F \exp(\beta^2 h(\boldsymbol{x}))d\boldsymbol{x} \tag{7.28}$$

where $h(\boldsymbol{x}) = l(\boldsymbol{x})/\beta_0^2$ *with* $\beta_0 = \sqrt{-\max_{\boldsymbol{x}\in F} l(\boldsymbol{x})}$.

Then we have as $\beta \to \infty$ *the following asymptotic relation*

$$\int_F e^{\beta^2 h(\boldsymbol{x})}d\boldsymbol{x} \sim (2\pi)^{(n-k)/2} \frac{e^{\beta^2 h(\boldsymbol{x}^*)}}{\prod_{i=1}^k \gamma_i \sqrt{\det(\boldsymbol{G})|\det(\boldsymbol{H}^*(\boldsymbol{x}^*))|}} \cdot \left(\frac{\beta_0}{\beta}\right)^{n+k}. \tag{7.29}$$

Here $\boldsymbol{G}$ *is the Gramian of the gradients* $\nabla g_1(\boldsymbol{x}^*), \ldots, \nabla g_k(\boldsymbol{x}^*)$.

PROOF: This theorem follows from theorem 48, p. 71. □

The quality of these approximations depends on the tail behavior of the p.d.f. of $g(\boldsymbol{X})$. If the tail has an exponential decay, the approximations should be good in general. If the deacy is very slow such approximations might give no sufficient results.

As said already in the last chapter these approximations can be used as basis for importance sampling methods. Such a concept is developed in [91]. In a benchmark study [56] it was found that for structural reliability problems this method was more efficient than the competing ones.

7.2.2 Examples

The following examples are from the article [29], Journal of the Engineering Mechanics Division, ASCE, Vol. 117 and are reprinted with permission.

EXAMPLE: Given are two independent χ^2-distributed random variables X_1 and X_2 with two degrees of freedom each having the p.d.f.

$$f_i(x_i) = \frac{1}{4} \cdot x_i \cdot e^{-x_i/2} \text{ for } x_i \geq 0. \tag{7.30}$$

The log likelihood function of the p.d.f. is then

$$l_i(x_i) = -\ln(4) + \ln(x_i) - x_i/2. \tag{7.31}$$

The sum of these random variables is again a χ^2 distributed random variable with four degrees of freedom. We take as limit state function $g(x_1, x_2)$

$$g(x_1, x_2) = 21 - x_1 - x_2. \tag{7.32}$$

The log likelihood function has only at the point $(\frac{21}{2}, \frac{21}{2})$ a global maximum. The derivatives of the function are

$$\frac{\partial l(\boldsymbol{x})}{\partial x_i} = \frac{1}{x_i} - \frac{1}{2}, \quad \frac{\partial^2 l(\boldsymbol{x})}{\partial x_i^2} = -\frac{1}{x_i^2}, \quad \frac{\partial^2 l(\boldsymbol{x})}{\partial x_1 \partial x_2} = 0. \tag{7.33}$$

The exact failure probability is given by

$$\mathrm{P}(X_1 + X_2 > 21) = 7.15 \times 10^{-3}. \tag{7.34}$$

The approximation with theorem 60, p. 110 gives instead

$$\mathrm{P}(X_1 + X_2 > 21) \sim 8.72 \times 10^{-3}. \tag{7.35}$$

□

EXAMPLE: This example is taken from [89], p. 78-79. In a similar form it appears also in [72] and [50]. Given is a two-dimensional random vector (X_1, X_2) with joint p.d.f.

$$f(x_1, x_2) = (x_1 + x_2 + x_1 x_2)\exp(-(x_1 + x_2 + x_1 x_2)) \quad \text{for} \quad x_1, x_2 \geq 0. \tag{7.36}$$

The log likelihood is then

$$l(x_1, x_2) = \ln(x_1 + x_2 + x_1 x_2) - x_1 - x_2 - x_1 x_2. \tag{7.37}$$

As limit state functions are taken

$$g_1(x_1, x_2) = 18 - 3x_1 - 2x_2, \tag{7.38}$$
$$g_2(x_1, x_2) = -x_2, \tag{7.39}$$
$$g_3(x_1, x_2) = -x_1. \tag{7.40}$$

The gradients are then

$$\nabla g_1(x_1, x_2) = \begin{pmatrix} -3 \\ -2 \end{pmatrix}, \tag{7.41}$$
$$\nabla g_2(x_1, x_2) = \begin{pmatrix} 0 \\ -1 \end{pmatrix}, \tag{7.42}$$
$$\nabla g_3(x_1, x_2) = \begin{pmatrix} -1 \\ 0 \end{pmatrix}. \tag{7.43}$$

To use theorem 61, p. 111 we need the point, where the log likelihood is maximal in the failure domain. There is only one such point $\boldsymbol{x}^0 = (6, 0)$. At the point $(0, 9)$ there is only a local maximum and not a global.

The gradient of the log likelihood is

$$\nabla l(x_1, x_2) = \begin{pmatrix} \dfrac{1 + x_2}{x_1 + x_2 + x_1 x_2} - 1 - x_2 \\ \\ \dfrac{1 + x_1}{x_1 + x_2 + x_1 x_2} - 1 - x_1 \end{pmatrix}. \tag{7.44}$$

At $\boldsymbol{x}^0 = (6, 0)$ the values are

$$\nabla l(6, 0) = \begin{pmatrix} \dfrac{1}{6} - 1 \\ \\ \dfrac{1+6}{6} - 1 - 6 \end{pmatrix} = \begin{pmatrix} -\dfrac{5}{6} \\ \\ -5\dfrac{5}{6} \end{pmatrix}. \tag{7.45}$$

For the density at $(6, 0)$ we get at this point

$$f(6, 0) = 6e^{-6} \approx 1.49 \times 10^{-2}. \tag{7.46}$$

At this point $g_1(6, 0) = g_2(6, 0) = 0$ and $g_3(6, 0) = -6 < 0$. Therefore only the limit state functions g_1 and g_2 are of interest here.

For the Gramian of the gradients of the limit state functions g_1 and g_2 at $(6, 0)$ we get

$$\det(\boldsymbol{A}^T \cdot \boldsymbol{A}) = \det \begin{pmatrix} 13 & 2 \\ 2 & 1 \end{pmatrix} = 9. \tag{7.47}$$

The coefficients γ_1 and γ_2 are

$$\gamma_1 = \frac{5}{18} \text{ and } \gamma_2 = 5\frac{5}{18}. \tag{7.48}$$

As approximation we obtain

$$\begin{aligned} \mathbb{P}(F) \quad &\sim \quad \frac{f(6, 0)}{\gamma_1 \cdot \gamma_2 \cdot \sqrt{\det(\boldsymbol{A}^T \boldsymbol{A})}} \\ = \frac{1.49 \times 10^{-2}}{1.47 \cdot 3} \quad &= \quad 3.39 \times 10^{-3}. \end{aligned} \tag{7.49}$$

The exact probability $\mathbb{P}(g_i(X_1, X_2) \le 0$ for $i = 1, 2, 3)$ is 2.94×10^{-3}.

In [89] are two different approximations obtained by the transformation method: 2.68×10^{-3} and 4.04×10^{-3}. The results are different, since if the Rosenblatt transformation is used, the order of the random variables can influence the result ([52]). But since this invariance is one of the basic requirements for useful algorithms, here is a further argument against the use of the transformation method in these cases.

To be quite correct, in [89] still a further approximation is constructed by combining the two linear approximations to a polyhedrical, which gives as result 2.94×10^{-3}, also the exact result. But since for this no generalization, which can be used in more complicated cases, is developed, it appears to be of limited value.

This example shows the essential advantage of the method of approximations in the original space. □

7.3 Sensitivity analysis

7.3.1 Parameter dependent densities

Often the exact value of the distribution parameters is unknown. A more general formulation of the reliability problem is then in the form

$$\mathbb{P}_{\boldsymbol{\vartheta},\boldsymbol{\tau}}(F) = \int\limits_{g(\boldsymbol{x},\boldsymbol{\tau})\leq 0} \exp(l(\boldsymbol{x},\boldsymbol{\vartheta}))\, d\boldsymbol{x}. \tag{7.50}$$

Here $l(\boldsymbol{x},\boldsymbol{\vartheta})$ is the log likelihood function. The two parameter vectors $\boldsymbol{\vartheta}$ and $\boldsymbol{\tau}$ describe the random influences.

In reliability problems often some of these parameters are design parameters which can be changed. For sensitivity analysis it is necessary to know the derivatives of the function $\mathbb{P}_{\boldsymbol{\vartheta},\boldsymbol{\tau}}(F)$ with respect to these parameters. A problem in connection with sensitivity analysis is what happens if some random variables are replaced by their mean to simplify the problem ([88]). The results of this section are based on [28].

Given is a family of n-dimensional probability densities $f(\boldsymbol{x},\boldsymbol{\vartheta})$, depending on a parameter $\boldsymbol{\vartheta} \in \mathbb{R}^k$. We assume that $f(\boldsymbol{x},\boldsymbol{\vartheta}) > 0$ for all $(\boldsymbol{x},\boldsymbol{\vartheta}) \in \mathbb{R}^n \times D$ with $D \subset \mathbb{R}^k$ being an open subset of $\mathbb{R}^k$. The log likelihood function $l(\boldsymbol{x},\boldsymbol{\vartheta}) = \ln(f(\boldsymbol{x},\boldsymbol{\vartheta}))$ is a function of $\boldsymbol{x}$ and $\boldsymbol{\vartheta}$.

For a fixed parameter value $\boldsymbol{\vartheta} \in D$ the probability content $\mathbb{P}_{\boldsymbol{\vartheta}}(F)$ of a subset $F \subset \mathbb{R}^n$ is given by

$$\mathbb{P}_{\boldsymbol{\vartheta}}(F) = \int\limits_F \exp(l(\boldsymbol{x},\boldsymbol{\vartheta}))\, d\boldsymbol{x}. \tag{7.51}$$

The partial derivative $\mathbb{P}_{\boldsymbol{\vartheta}}(F))$ with respect to ϑ_i is found if the necessary conditions are fulfilled, by interchanging differentiation and integration

$$\frac{\partial \mathbb{P}_{\boldsymbol{\vartheta}}(F)}{\partial \vartheta_i} = \int\limits_F \frac{\partial l(\boldsymbol{x},\boldsymbol{\vartheta})}{\partial \vartheta_i} \exp(l(\boldsymbol{x},\boldsymbol{\vartheta}))\, d\boldsymbol{x}. \tag{7.52}$$

The partial elasticity $\epsilon_i(I\!P_{\boldsymbol{\vartheta}}(F))$ of the probability $I\!P_{\boldsymbol{\vartheta}}(F)$ with respect to ϑ_i is then

$$\epsilon_i(I\!P_{\boldsymbol{\vartheta}}(F)) = \frac{\vartheta_i}{I\!P_{\boldsymbol{\vartheta}}(F)} \int_F \frac{\partial l(\boldsymbol{x}, \boldsymbol{\vartheta})}{\partial \vartheta_i} \exp(l(\boldsymbol{x}, \boldsymbol{\vartheta}))\, d\boldsymbol{x}. \tag{7.53}$$

As scaling factor a suitable β_0 is chosen, for example $\beta_0 = \sqrt{-\max_F l(\boldsymbol{x}, \boldsymbol{\vartheta}^*)}$ for a $\boldsymbol{\vartheta}^*$ with $\max_F l(\boldsymbol{x}, \boldsymbol{\vartheta}^*) < 0$.

Replacing now in the integral the function $l(\boldsymbol{x}, \boldsymbol{\vartheta})$ by $h(\boldsymbol{x}, \boldsymbol{\vartheta}) = l(\boldsymbol{x}, \boldsymbol{\vartheta})/\beta_0^2$, we obtain

$$\frac{\partial I\!P_{\boldsymbol{\vartheta}}(F)}{\partial \vartheta_i} = \int_F \beta_0^2 \frac{\partial h(\boldsymbol{x}, \boldsymbol{\vartheta})}{\partial \vartheta_i} \exp(\beta_0^2 h(\boldsymbol{x}, \boldsymbol{\vartheta}))\, d\boldsymbol{x} \tag{7.54}$$

and in the same way for the partial elasticity with respect to ϑ_i

$$\epsilon_i(I\!P_{\boldsymbol{\vartheta}}(F)) = \frac{\vartheta_i}{I\!P_{\boldsymbol{\vartheta}}(F)} \int_F \beta_0^2 \frac{\partial h(\boldsymbol{x}, \boldsymbol{\vartheta})}{\partial \vartheta_i} \exp(\beta_0^2 h(\boldsymbol{x}, \boldsymbol{\vartheta}))\, d\boldsymbol{x}. \tag{7.55}$$

We define now the integrals $I\!P^{\beta}_{\boldsymbol{\vartheta}}(F)$ by

$$I\!P^{\beta}_{\boldsymbol{\vartheta}}(F) = \int_F \exp(\beta^2 h(\boldsymbol{x}, \boldsymbol{\vartheta}))\, d\boldsymbol{x}. \tag{7.56}$$

We assume now that there is only one point $\boldsymbol{x}^*$ on the boundary of F, where the log likehood function achieves its global maximum with respect to F. Then using theorem 60, p. 110 we get for $I\!P^{\beta}_{\boldsymbol{\vartheta}}(F)$ as $\beta \to \infty$ the asymptotic approximation

$$I\!P^{\beta}_{\boldsymbol{\vartheta}}(F) \sim (2\pi)^{(n-1)/2} \frac{e^{\beta^2 h(\boldsymbol{x}^*)}}{|\nabla l(\boldsymbol{x}^*)| \sqrt{|\det(\boldsymbol{A}^T(\boldsymbol{x}^*)\boldsymbol{H}(\boldsymbol{x}^*)\boldsymbol{A}(\boldsymbol{x}^*))|}} \left(\frac{\beta_0}{\beta}\right)^{n+1}. \tag{7.57}$$

Here the functions and matrices are defined as in theorem 60.

For the partial derivatives and elasticities then the following asymptotic approximations are obtained using theorem 46, p. 67

$$\frac{\partial I\!P^{\beta}_{\boldsymbol{\vartheta}}(F)}{\partial \vartheta_i} \sim \frac{\partial l(\boldsymbol{x}^*, \boldsymbol{\vartheta})}{\partial \vartheta_i} I\!P^{\beta}_{\boldsymbol{\vartheta}}(F) \left(\frac{\beta}{\beta_0}\right)^2, \quad \beta \to \infty, \tag{7.58}$$

$$\epsilon_i(I\!P^{\beta}_{\boldsymbol{\vartheta}}(F)) \sim \vartheta_i \cdot \frac{\partial l(\boldsymbol{x}^*, \boldsymbol{\vartheta})}{\partial \vartheta_i} \left(\frac{\beta}{\beta_0}\right)^2, \quad \beta \to \infty. \tag{7.59}$$

This yields for the partial derivative with respect to ϑ_i at $\boldsymbol{\vartheta}^*$

$$\left.\frac{\partial I\!P^{\beta_0}_{\boldsymbol{\vartheta}}(F)}{\partial \vartheta_i}\right|_{\boldsymbol{\vartheta}=\boldsymbol{\vartheta}^*} \sim \frac{\partial l(\boldsymbol{x}^*, \boldsymbol{\vartheta}^*)}{\partial \vartheta_i} I\!P^{\beta_0}_{\boldsymbol{\vartheta}}(F). \tag{7.60}$$

The general case, where also the limit state function depends on the parameter vector can be treated in a similar way. We restrict our considerations to the case of one parameter τ.

We define the integrals $\mathbb{P}^{\beta}_{\tau}(F)$ by

$$\mathbb{P}^{\beta}_{\tau}(F) = \int\limits_{g(\boldsymbol{x},\tau)\leq 0} \exp(\beta^2 h(\boldsymbol{x},\tau))\, d\boldsymbol{x}. \tag{7.61}$$

with $h(\boldsymbol{x},\tau) = \beta_0^{-2} l(\boldsymbol{x},\tau)$.

From theorem 21, p. 23 we get the following representation of the partial derivative with respect to the variable τ

$$\begin{aligned} \frac{\partial \mathbb{P}^{\beta}_{\tau}(F)}{\partial \tau} &= \beta^2 \int\limits_{g(\boldsymbol{x},\tau)\leq 0} \frac{\partial h(\boldsymbol{x},\tau)}{\partial \tau} e^{\beta^2 h(\boldsymbol{x},\tau)}\, d\boldsymbol{x} \\ &- \int\limits_{g(\boldsymbol{x},\tau)=0} \frac{g_\tau(\boldsymbol{x},\tau)}{|\nabla g_{\boldsymbol{x}}(\boldsymbol{x},\tau)|} e^{\beta^2 h(\boldsymbol{x},\tau)}\, ds_G(\boldsymbol{x}). \end{aligned} \tag{7.62}$$

with $ds_G(\boldsymbol{x})$ denoting surface integration over G.

For these integrals asymptotic approximations can be derived using the results in chapter 4 (theorem 46, p. 67 for the domain integral and theorem 47, p. 70 for the surface integral). For the first integral the approximation is

$$\begin{aligned} \beta^2 \int\limits_{g(\boldsymbol{x},\tau)\leq 0} \frac{\partial h(\boldsymbol{x},\tau)}{\partial \tau} e^{\beta^2 h(\boldsymbol{x},\tau)} d\boldsymbol{x} &\sim \beta^2 \frac{\partial h(\boldsymbol{x}^*,\tau)}{\partial \tau} \int\limits_{g(\boldsymbol{x},\tau)\leq 0} e^{\beta^2 h(\boldsymbol{x},\tau)} d\boldsymbol{x} \\ &\sim \left(\frac{\beta}{\beta_0}\right)^2 \frac{\partial l(\boldsymbol{x}^*,\tau)}{\partial \tau} \mathbb{P}^{\beta}_{\tau}(F). \end{aligned} \tag{7.63}$$

For the second integral, the surface integral, we get

$$\begin{aligned} &\int\limits_{g(\boldsymbol{x},\tau)=0} \frac{g_\tau(\boldsymbol{x},\tau)}{|\nabla g_{\boldsymbol{x}}(\boldsymbol{x},\tau)|} e^{\beta^2 h(\boldsymbol{x},\tau)}\, ds_G(\boldsymbol{x}) \\ \sim\; &\frac{g_\tau(\boldsymbol{x}^*,\tau)}{|\nabla g_{\boldsymbol{x}}(\boldsymbol{x}^*,\tau)|} \int\limits_{g(\boldsymbol{x},\tau)=0} e^{\beta^2 h(\boldsymbol{x},\tau)} ds_G(\boldsymbol{x}) \\ \sim\; &\left(\frac{\beta}{\beta_0}\right)^2 \frac{g_\tau(\boldsymbol{x}^*,\tau)|\nabla_{\boldsymbol{x}} l(\boldsymbol{x}^*,\tau)|}{|\nabla g_{\boldsymbol{x}}(\boldsymbol{x}^*,\tau)|} \mathbb{P}^{\beta}_{\tau}(F). \end{aligned} \tag{7.64}$$

The last relation in the equation above is found by comparing the asymptotic approximations in the theorems 46, p. 67 and 47, p. 70

$$\int\limits_{g(\boldsymbol{x},\tau)=0} e^{\beta^2 h(\boldsymbol{x},\tau)}\, ds_G(\boldsymbol{x}) \sim \left(\frac{\beta}{\beta_0}\right)^2 |\nabla_{\boldsymbol{x}} l(\boldsymbol{x}^*,\tau)| \int\limits_{g(\boldsymbol{x},\tau)\leq 0} e^{\beta^2 h(\boldsymbol{x},\tau)}\, d\boldsymbol{x}. \tag{7.65}$$

Multiplied by $\left(\frac{\beta}{\beta_0}\right)^2 |\nabla_{\boldsymbol{x}} l(\boldsymbol{x}^*, \tau)|$ the surface integral is an asymptotic approximation for the probability $\mathbb{P}_\tau^\beta(F)$.

Adding the two approximations gives the final result

$$\frac{\partial \mathbb{P}_\tau^\beta(F)}{\partial \tau} \sim \left(\frac{\partial l(\boldsymbol{x}^*, \tau)}{\partial \tau} - \frac{g_\tau(\boldsymbol{x}^*, \tau)|\nabla_{\boldsymbol{x}} l(\boldsymbol{x}^*, \tau)|}{|\nabla g_{\boldsymbol{x}}(\boldsymbol{x}^*, \tau)|}\right) \mathbb{P}_\tau^\beta(F) \left(\frac{\beta}{\beta_0}\right)^2. \tag{7.66}$$

An application of these results in structural reliability can be found in [41].

7.3.2 Example

Let be given two independent random variables X_1 and X_2, each with a lognormal distribution with $\mathbb{E}(\ln(X_i)) = \mu$ and $\mathrm{var}(\ln(X_i)) = \sigma^2$ for $i = 1, 2$. The joint probability density function $f(x_1, x_2)$ of these random variables is then

$$f(x_1, x_2) = \frac{1}{2\pi\sigma^2 x_1 x_2} \exp\left(-\frac{\sum_{i=1}^2 (\ln(x_i) - \mu)^2}{2\sigma^2}\right). \tag{7.67}$$

The log likelihood function is

$$l(x_1, x_2) = -\ln(2\pi) - 2\ln(\sigma) - \ln(x_1) - \ln(x_2) - \frac{\sum_{i=1}^2 (\ln(x_i) - \mu)^2}{2\sigma^2}. \tag{7.68}$$

We consider as limit state function the function

$$g_\gamma(x_1, x_2) = \gamma^2 - x_1 \cdot x_2. \tag{7.69}$$

For the derivatives of the log likelihood function we get

$$\begin{aligned} \frac{\partial l(x_1, x_2)}{\partial x_i} &= -x_i^{-1}\left(1 + \frac{\ln(x_i) - \mu}{\sigma^2}\right) & (7.70) \\ \frac{\partial^2 l(x_1, x_2)}{\partial x_i^2} &= x_i^{-2}\left(1 + \frac{\ln(x_i) - \mu}{\sigma^2} - \sigma^{-2}\right). & (7.71) \end{aligned}$$

The gradients of l and g are with $c_\gamma = (1 + \frac{\ln(\gamma) - \mu}{\sigma^2})$

$$\nabla l(x_1, x_2) = \begin{pmatrix} -x_1^{-1} c_\gamma \\ -x_2^{-1} c_\gamma \end{pmatrix} \text{ and } \nabla g(x_1, x_2) = \begin{pmatrix} -x_2 \\ -x_1 \end{pmatrix}. \tag{7.72}$$

At the point (γ, γ), the only global maximum of the log likelihood function, we obtain for the norm of these vectors

$$\begin{aligned} |\nabla l(\gamma, \gamma)| &= \sqrt{2}\gamma^{-1} c_\gamma, & (7.73) \\ |\nabla g(\gamma, \gamma)| &= \sqrt{2}\gamma. \end{aligned}$$

For the Hessian at (γ, γ)

$$\boldsymbol{H}_{\tilde{j}} = \left(l_{ij}(\gamma, \gamma) - \frac{|\nabla l(\gamma, \gamma)|}{|\nabla g(\gamma, \gamma)|} g_{ij}(\gamma, \gamma) \right)_{i,j=1,2} \tag{7.74}$$

we get then

$$\boldsymbol{H}_{\tilde{j}} = \begin{pmatrix} \gamma^{-2}(c_\gamma - \sigma^{-2}) & c_\gamma \gamma^{-2} \\ c_\gamma \gamma^{-2} & \gamma^{-2}(c_\gamma - \sigma^{-2}) \end{pmatrix}. \tag{7.75}$$

The gradient of l at (γ, γ) is parallel to the vector $(1, 1)$, therefore a unit vector orthogonal to this vector is the vector $\boldsymbol{a}_1 = (1/\sqrt{2}, -1/\sqrt{2})$. This gives

$$\sqrt{|\boldsymbol{a}_1^T \boldsymbol{H}_{\tilde{j}} \boldsymbol{a}_1|} = (\gamma\sigma)^{-1}. \tag{7.76}$$

The joint density at (γ, γ) is

$$f(\gamma, \gamma) = \frac{1}{2\pi(\gamma\sigma)^2} \exp\left(-\frac{(\ln(\gamma) - \mu)^2}{\sigma^2}\right). \tag{7.77}$$

Now, using the approximation formula (7.22), we obtain as approximation for the failure probability the following

$$\begin{aligned} P_A(F) &= \frac{\sqrt{2\pi}}{\sqrt{2}(1 + \frac{\ln(\gamma) - \mu}{\sigma^2}) \cdot \gamma^{-2}\sigma^{-1}} \cdot f(\gamma, \gamma) \qquad (7.78) \\ &= \frac{1}{\sqrt{2\pi}\sqrt{2}\sigma(1 + \frac{\ln(\gamma) - \mu}{\sigma^2})} \exp\left(-\frac{(\ln(\gamma) - \mu)^2}{\sigma^2}\right). \end{aligned}$$

The partial derivatives of the log likelihood function at (γ, γ) with respect to the parameters μ and σ are

$$\frac{\partial l}{\partial \mu} = 2 \cdot \frac{\ln(\gamma) - \mu}{\sigma^2}, \tag{7.79}$$

$$\frac{\partial l}{\partial \sigma} = 2 \cdot (-\sigma^{-1} + \sigma^{-3}(\ln(\gamma) - \mu)^2). \tag{7.80}$$

For the derivative of g at (γ, γ) with respect to γ we find

$$g_\gamma = 2\gamma. \tag{7.81}$$

The exact probability content of the failure domain is given by

$$P(F) = \Phi\left(-\sqrt{2} \cdot \frac{\ln(\gamma) - \mu}{\sigma}\right). \tag{7.82}$$

By differentiating this function with respect to the parameters μ, σ and γ, we obtain the true sensitivities of the failure probability to changes in these parameters.

For the partial elasticities of the approximations $\mathbb{P}_A(F)$ we get using equation (7.66) that

$$\epsilon_\mu(P_A(F)) = 2\mu\frac{\ln(\gamma)-\mu}{\sigma^2}, \tag{7.83}$$

$$\epsilon_\sigma(P_A(F)) = 2\left(-1+\left(\frac{\ln(\gamma)-\mu}{\sigma}\right)^2\right), \tag{7.84}$$

$$\epsilon_\gamma(P_A(F)) = 2\left(1+\frac{\ln(\gamma)-\mu}{\sigma^2}\right). \tag{7.85}$$

The partial elasticities of the exact failure probability are asymptotically for $\gamma \to \infty$

$$\epsilon_\mu(P(F)) \sim 2\mu\cdot\frac{\ln(\gamma)-\mu}{\sigma^2}, \tag{7.86}$$

$$\epsilon_\sigma(P(F)) \sim 2\cdot\left(\frac{\ln(\gamma)-\mu}{\sigma}\right)^2, \tag{7.87}$$

$$\epsilon_\gamma(P(F)) \sim 2\cdot\frac{\ln(\gamma)-\mu}{\sigma^2}. \tag{7.88}$$

Asymptotically the approximations of the elasticities approach the true elasticities.

Chapter 8

Crossing Rates of Stochastic Processes

8.1 Introduction

In many problems of reliability analysis it is necessary to consider time variant influences modelled by stochastic processes. Here the extreme value distribution of the process is of interest. Depending on the structure of the process there are different possibilities to calculate this distribution.

For Markov processes it is possible by using these Markov properties to find solutions. For stochastic processes, which are differentiable and therefore not Markovian, other methods must be used.

Textbooks which treat also vector processes are [101] and [108]. The problem of extreme value distributions of differentiable processes is studied in [43] and [82].

8.2 Definition and Properties of Stochastic Processes

Definition 22 *Given is a probability space* $(\Omega, \mathcal{B}, \mathbb{P})$ *and a parameter set* T. *A stochastic process is defined as a real function* $x(t, \omega)$ *which for each* $t \in T$ *is a measurable function of* $\omega \in \Omega$.

Normally T is a time interval and $x(t)$ is the value of the process at time t. In the following an one-dimensional stochastic process is denoted by $x(t)$.

Definition 23 *Given is a probability space* $(\Omega, \mathcal{B}, \mathbb{P})$ *and a parameter set* T. *A stochastic vector process is defined as a function* $\mathbf{x}(t, \omega) : T \times \Omega \to \mathbb{R}^n, (t, \omega) \mapsto \mathbf{x}(T, \omega)$ *which for each* $t \in T$ *is a measurable function of* $\omega \in \Omega$.

In the following an n-dimensional vector process $(x_1(t), \ldots, x_n(t))$ is denoted by $\mathbf{x}(t)$.

Definition 24 *The autocorrelation matrix* $\boldsymbol{R}(t_1, t_2)$ *of a vector process* $\boldsymbol{x}(t) = (x_1(t), \ldots, x_n(t))$ *is given by*

$$\boldsymbol{R}(t_1, t_2) = (\mathbb{E}(x_i(t_1)x_j(t_2)))_{i,j=1,\ldots,n} \tag{8.1}$$

and its autocovariance matrix $\boldsymbol{C}(t_1, t_2)$ *is given by*

$$\boldsymbol{C}(t_1, t_2) = (\mathbb{E}(x_i(t_1)x_j(t_2)) - \mathbb{E}(x_i(t_1)\mathbb{E}(x_j(t_2)))_{i,j=1,\ldots,n}. \tag{8.2}$$

Definition 25 *The cross-correlation matrix* $\boldsymbol{R_{xy}}(t_1, t_2)$ *of two n-dimensional vector processes* $\boldsymbol{x}(t)$ *and* $\boldsymbol{y}(t)$ *is given by*

$$\boldsymbol{R_{xy}}(t_1, t_2) = (\mathbb{E}(x_i(t_1)y_j(t_2)))_{i,j=1,\ldots,n} \tag{8.3}$$

and their cross-covariance matrix $\boldsymbol{C_{xy}}(t_1, t_2)$ *is given by*

$$\boldsymbol{C_{xy}}(t_1, t_2) = (\mathbb{E}(x_i(t_1)y_j(t_2)) - \mathbb{E}(x_i(t_1))\mathbb{E}(y_j(t_2))))_{i,j=1,\ldots,n}. \tag{8.4}$$

Definition 26 *A vector process* $\boldsymbol{x}(t)$ *is called wide sense stationary if its mean value vector*

$$\mathbb{E}(\boldsymbol{x}(t)) = \boldsymbol{\mu} \tag{8.5}$$

is equal to a constant vector $\boldsymbol{\mu}$ *and the autocorrelation matrix depends only on the difference* $t_1 - t_2$

$$\boldsymbol{R}(t_1, t_2) = \boldsymbol{R}(t_1 + \tau, t_2 + \tau) \text{ for any } \tau \in \mathbb{R}. \tag{8.6}$$

Definition 27 *A vector process* $\boldsymbol{x}(t)$ *is called strictly stationary if the two processes processes* $\boldsymbol{x}(t)$ *and* $\boldsymbol{x}(t+\tau)$ *have the same finite dimensional distributions for any* $\tau \in \mathbb{R}$, *i.e. for all* $\tau \in \mathbb{R}$, *all* $k \in \mathbb{N}$ *and all vectors* $\boldsymbol{z}_1, \ldots, \boldsymbol{z}_k$

$$\mathbb{P}(\boldsymbol{x}(t_1) \le \boldsymbol{z}_1, \ldots, \boldsymbol{x}(t_k) \le \boldsymbol{z}_k) = \mathbb{P}(\boldsymbol{x}(t_1 + \tau) \le \boldsymbol{z}_1, \ldots, \boldsymbol{x}(t_k + \tau) \le \boldsymbol{z}_k). \tag{8.7}$$

Definition 28 *A stochastic process is said to be a normal process if all finite dimensional distributions of the process are normal distributions.*

A normal process is completely determined in terms of its means and autocovariances. Therefore a wide sense stationary normal process is also a strictly stationary process.

Definition 29 *A stationary stochastic process* $x(t)$ *is differentiable in the m.s. (mean square) sense if there exists a stationary stochastic process* $x'(t)$ *such that for any* t

$$\lim_{h \to 0} \mathbb{E}\left[\left(\frac{x(t+h) - x(t)}{h} - x'(t)\right)^2\right] = 0. \tag{8.8}$$

Theorem 62 *A stationary stochastic process* $x(t)$ *is differentiable in the m.s. sense if its autocorrelation function* $r(t)$ *has derivatives up to order two.*

PROOF: A proof is given in [43], p. 84. □

Slightly stronger is the differentiability with probability one.

Definition 30 *A stationary stochastic process $x(t)$ is said to be differentiable with probability one if with probability one each path $x(t,\omega)$ has a derivative* $\frac{\partial x(t,\omega)}{\partial t}$.

The difference between these two definitions is outlined in [43], p. 84/5.

Let $x(t)$ be a stationary normal process differentiable with probability one. Then its derivative process $x'(t)$ is again a stationary normal process (see [43], chapter 7.8).

Assume now that $x(t)$ is a stationary normal process differentiable with probability one and such that

$$\mathbb{E}(x(t)) = 0 \text{ and } \operatorname{var}(x(t)) = 1.$$

Its autocorrelation function is $r(t)$. Then the variances and covariances of the derivative process are (see [43], p. 177)

$$\begin{aligned} \operatorname{var}(x'(t)) &= -r''(0), \\ \operatorname{cov}(x(0), x'(t)) &= r'(t). \end{aligned} \tag{8.9}$$

Especially we have $\operatorname{cov}(x(t), x'(t)) = r'(0) = 0$. (Since $r(t)$ is an even function, its derivative at zero must be zero.) Since the random variables $x(t)$ and $x'(t)$ are both normally distributed, from this follows that they are independent too.

Definition 31 *A stationary stochastic vector process is said to be a normal process if all finite-dimensional distributions are normal distributions.*

Given is a stationary normal vector process $\boldsymbol{x}(t) = (x_1(t), \ldots, x_n(t))$. Assume that the process is centered and standardized, i.e.

$$\mathbb{E}(x_i(t)) = 0 \text{ for } i = 1, \ldots, n, \tag{8.10}$$

$$\operatorname{cov}(x_i(t), x_j(t)) = \delta_{ij} \text{ for } i, j = 1, \ldots, n. \tag{8.11}$$

Assume further that the process has continuously differentiable paths with probability one. The derivative process $\boldsymbol{x}'(t) = (x_1'(t), \ldots, x_n'(t))$ exists therefore and is again a normal process. The autocovariance matrix of $\boldsymbol{x}'(t)$ is then

$$(\operatorname{cov}(x_i'(0), x_j'(t)))_{i,j=1,\ldots,n} = (-r_{ij}''(t))_{i,j=1,\ldots,n}. \tag{8.12}$$

The cross-covariance matrix between the process and the derivative process is (see [108], p. 317).

$$(r_{ij}'(t))_{i,j=1,\ldots,n} = (\operatorname{cov}(x_i(0), x_j'(t)))_{i,j=1,\ldots,n} = (-\operatorname{cov}(x_i'(0), x_j(t)))_{i,j=1,\ldots,n}. \tag{8.13}$$

This gives for $t = 0$

$$\begin{aligned} (r_{ij}'(0))_{i,j=1,\ldots,n} &= (\operatorname{cov}(x_i(0), x_j'(0)))_{i,j=1,\ldots,n} \\ = (-\operatorname{cov}(x_i'(0), x_j(0)))_{i,j=1,\ldots,n} &= (-r_{ji}'(0))_{i,j=1,\ldots,n}. \end{aligned} \tag{8.14}$$

The matrix $(r_{ij}'(0))_{i,j=1,\ldots,n}$ is therefore skew-symmetric and so $x_i(t)$ and $x_i'(t)$ are uncorrelated. In the case of a normal process they are independent.

We write for the autocorrelation matrix of the process $\boldsymbol{R}(t)$. This gives for the cross-correlation matrix $\boldsymbol{R}_{\boldsymbol{x},\boldsymbol{x}'}(0,t)$ between the process and the derivative process

$$\boldsymbol{R}_{\boldsymbol{x},\boldsymbol{x}'}(0,t) = \boldsymbol{R}'(t). \tag{8.15}$$

and for the autocorrelation matrix $\boldsymbol{R}_{\boldsymbol{x}'}(0,t)$ of the derivative processess

$$\boldsymbol{R}_{\boldsymbol{x}'}(0,t) = -\boldsymbol{R}''(t). \tag{8.16}$$

By a suitable linear transformation (see lemma 6, p. 11) it can be achieved always that the covariance matrix of the derivative process is a diagonal matrix. Then we have for the two covariance matrices of the process and the derivative process

$$\boldsymbol{R}(0) = \boldsymbol{I}_n \ , \ \boldsymbol{R}''(0) = \begin{pmatrix} -\tilde{\sigma}_1^2 & \dots & 0 \\ 0 & \ddots & 0 \\ 0 & \dots & -\tilde{\sigma}_n^2 \end{pmatrix}. \tag{8.17}$$

Here $\tilde{\sigma}_i^2 = \text{var}(x_i'^2(t))$.

This gives a simple structure for the process. Only the form of the cross-correlation matrix $\boldsymbol{R}'(0)$ between the process and the derivative process is not in a standardized form.

8.3 Maxima and Crossings of a Stochastic Process

Given is a strictly stationary stochastic process $x(t)$ with continuous time parameter t. In the following we assume that the distribution function $F(x) = \mathbb{P}(x(t) \le x)$ is continuous and that with probability one all paths are continuous functions. Conditions which ensure that these assumptions are fulfilled are given in [43].

The problem, which we will consider in the following, is the extreme value distribution of the process $x(t)$. We want to find the distribution function of the random variable

$$M(T) = \sup\{x(t); 0 \le t \le T\}. \tag{8.18}$$

In general it is not possible to calculate the exact form of this function. Therefore we attempt to find approximations. If the paths of the process are continuously differentiable with probability one, this property can be used to derive such approximations.

This topic was studied in the book of Cramer and Leadbetter [43] and then in Leadbetter, Lindgren and Rootzén [82]. Solutions in an analytic form are available almost only for normal processes. In [22] the author has derived approximations for the maxima of filtered Poisson processes converging towards a normal process. Approaches for a more general theory can be found in [2] and [3].

The use of such differentiable vector processes for the reliability of structures is shown in [137] and [17]. Applications in ship construction are given in [60].

To find an approximation for the extreme value distribution of a stochastic process, approximations for the probability that the process crosses a level in some small time interval are derived. In this subsection the meaning of "crossing a level" will be defined in a precise way (see [82], p. 146).

Definition 32 *With G_u we denote the set of all functions $f : T \to \mathbb{R}$ which are continuous on T and not identically to u in any subinterval.*

If $x(t)$ is a stationary process with the properties described in the last subsection, its sample paths are with probability one members of G_u, since

$$\mathbb{P}(x(t) \notin G_u) \leq \sum_{j=1}^{\infty} \mathbb{P}(x(t_j) = u),$$

where $\{t_j\}$ is an enumeration of the rational points in T. Due to the continuity of $F(x)$ we have $\mathbb{P}(x(t_j) = u) = 0$ for all t_j.

Now we consider the functions f in G_u.

Definition 33 *A function $f \in G_u$ has at the point $t_0 \in T$ a strict u-upcrossing if for some $\epsilon > 0$*

$$f(t) \leq u \text{ in } (t_0 - \epsilon, t_0) \text{ and } f(t) > u \text{ in } (t_0, t_0 + \epsilon). \tag{8.19}$$

Analogous a strict u-downcrossings is defined if there is some $\epsilon > 0$ with $f(t) > u$ in $(t_0 - \epsilon, t_0)$ and $f(t) < u$ for $(t_0, t_0 + \epsilon)$.

Definition 34 *A function $f \in G_u$ has at a point $t_0 \in T$ a u-upcrossing if for some $\epsilon > 0$ and all $\eta > 0$ $f(t) \leq u$ for all $t \in (t_0 - \epsilon, t_0)$ and $f(t) > u$ for some $t \in (t_0, t_0 + \eta)$.*

In the same way u-downcrossings are defined.

Definition 35 *A function $f \in G_u$ has at a point $t_0 \in T$ a (strict) u-crossing if it has there either a (strict) u-upcrossing or a (strict) u-downcrossing.*

We write $U(T)$ to denote the number of u-upcrossings in the interval $(0, T)$

$$U(T) = \#\{t \in (0, T); u\text{-upcrossing in } t\}. \tag{8.20}$$

These upcrossings can be used to obtain approximations for the extreme value distribution of the stochastic process if the process is smooth enough. We have the relation

$$\begin{aligned} \mathbb{P}(U(T) = 0) &= \mathbb{P}(x(0) < u, U(T) = 0) + \mathbb{P}(x(0) \geq u, U(T) = 0) \quad (8.21) \\ &= \mathbb{P}(\max_{0 \leq t \leq T} x(t) < u) + \mathbb{P}(x(0) \geq u, U(T) = 0). \end{aligned}$$

If we rewrite this equation we get

$$\mathbb{P}(\max_{0\leq t\leq T} x(t) < u) = \mathbb{P}(U(T) = 0) - \mathbb{P}(x(0) \geq u, U(T) = 0). \tag{8.22}$$

On the other hand we get then that

$$\mathbb{P}(\max_{0\leq t\leq T} x(t) > u) = \mathbb{P}(x(0) > u, U(T) = 0) + \mathbb{P}(x(0) \leq u, U(T) > 0). \tag{8.23}$$

Since $U(T)$ is a non-negative, integer-valued random variable we find

$$\mathbb{P}(U(T) > 0) = \sum_{i=1}^{\infty} \mathbb{P}(U(T) = i) \leq \sum_{i=1}^{\infty} i\mathbb{P}(U(T) = i) = \mathbb{E}(U(T)). \tag{8.24}$$

And this gives

$$\mathbb{P}(\max_{0\leq t\leq T} x(t) > u) \leq \mathbb{E}(U(T)) + \mathbb{P}(x(0) > u). \tag{8.25}$$

For a differentiable process the number of crossings of a level in a finite time interval is in general finite with probability one. This gives therefore a simple upper bound for the extreme value distribution.

The conditions, when for a differentiable process the crossing rate over a level is given by the conditional expectation of the derivative process was first derived by E.V. Bulinskaya [37]. The following lemma is a simplified form of the theorem 2 in [81], formulated for stationary processes.

We consider a stationary stochastic process $Y(t)$ on the time interval $[0,1]$. Let $g_\tau(x,z)$ denote the joint p.d.f. of $Y(t)$ and $(Y(t+\tau) - Y(t))/\tau$. Under some regularity conditions the second random variable will converge towards the derivative $Y'(t)$ of the process.

Lemma 63 *Suppose that* $Y(t)$ *has continuous sample paths with probability one and that*

1. $g_\tau(x,z)$ *is continuous in* x *for each* z *and* τ,
2. $g_\tau(x,z) \to p(x,z)$ *as* $\tau \to 0$ *uniformly in* x *for each* z,
3. $g_\tau(x,z) \leq h(z)$ *for all* τ, x, *where* $\int_{\mathbb{R}} |z|h(z)\,dz < \infty$.

Then

$$E(C(u)) = \int_{\mathbb{R}} |z|p(u,z)\,dz. \tag{8.26}$$

PROOF: See the article by Leadbetter [81], theorem 2. □

In [94] the conditions under which such results, i.e. the representation of crossing rates by the joint conditional expectation of the derivative process, are worked out in detail.

If we consider an n-dimensional vector process $\boldsymbol{x}(t)$ and a function $g : \mathbb{R}^n \to \mathbb{R}$, we get an one-dimensional process $g(\boldsymbol{x}(t))$. A u-crossing of the process $g(\boldsymbol{x}(t))$ corresponds to a crossing through the hypersurface $\{\boldsymbol{x}; g(\boldsymbol{x}) = u\}$ of the process $\boldsymbol{x}(t)$.

In the case of normal processes it can be shown in many cases that as $u \to \infty$ for a suitable scaling of the point process of crossings this point process converges towards a homogeneous Poisson point process. For stationary normal processes with independent components such results are proved in [85].

For stationary processes the expected number of u-upcrossings is under some regularity conditions equal to 1/2 of the expected number of the u-crossings. Therefore the approximations for the number of crossings give also approximations for the number of upcrossings. Extreme value distributions of functions of stochastic processes are of interest in a number of problems in reliability theory (see for example [17]). We consider here only stationary processes, but these methods are applicable also for non-stationary processes. Non-stationary processes are used in earthquake modelling. First results for such processes are given in in [27].

8.4 Crossings through an Hypersurface

In the following we consider only stationary normal processes. In this section asymptotic approximations for the expected number $\mathbb{E}(C(G))$ of crossings through an hypersurface G in $\mathbb{R}^n$ are derived.

The following result gives a formula for calculating the expected number of crossings of a differentiable normal vector process through a hypersurface. Belyaev[8] proved 1968 a theorem that the expected number of crossings of a differentiable vector process through an hypersurface is given a surface integral over this surface under some regularity conditions.

We consider a stationary continuously differentiable normal process $\boldsymbol{x}(t) = (x_1(t), \dots, x_n(t))$ with derivative process $\boldsymbol{x}'(t) = (x_1'(t), \dots, x_n'(t))$, which is standardized and centered. Further is given by a continuously differentiable function $g : \mathbb{R}^n \to \mathbb{R}$ a C^1-manifold $G = \{\boldsymbol{x}; g(\boldsymbol{x}) = 0\}$. Here it is assumed that $\nabla g(\boldsymbol{x}) \neq \mathbf{o}$ for all $\boldsymbol{x} \in G$. Then under some regularity conditions the expected number of crossings of the process $\boldsymbol{x}(t)$ through the hypersurface G during $(0, 1)$ is given by

$$\mathbb{E}(C(G)) = \int_G \mathbb{E}(|\boldsymbol{n}^T(\boldsymbol{x})\boldsymbol{x}'(t)|; \boldsymbol{x}(t) = \boldsymbol{x})\varphi_n(\boldsymbol{x})ds_G(\boldsymbol{x}). \tag{8.27}$$

Here $\boldsymbol{n}(\boldsymbol{x}) = |\nabla g(\boldsymbol{x})|^{-1}\nabla g(\boldsymbol{x})$ is the surface normal at $\boldsymbol{x}$ and $ds_G(\boldsymbol{x})$ denotes surface integration. over G. $\mathbb{E}(X; Y = y)$ denotes the conditional expectation of the random variable X under the condition $Y = y$.

To explain the equation above: The last factor $\varphi_n(\boldsymbol{x})$ approximates the probability that the process is at the point $\boldsymbol{x}$ and the first factor $\mathbb{E}(|\boldsymbol{n}^T(\boldsymbol{x})\boldsymbol{x}'(t)|; \boldsymbol{x}(t) = \boldsymbol{x})$ gives the conditional expectation of the absolute value of the projection of the derivative process orthogonal to the surface. The product is the probability flow through the surface at this point.

Theorem 64 *Let $x(t)$ be an one-dimensional stationary normal process with $var(x(t)) = 1$ whose autocorrelation function $r(t)$ is twice continuously differentiable. Then the expected value $\mathbb{E}(C(u))$ of the number of u-crossings in the time interval $(0,1)$ is finite for all $u \in \mathbb{R}$ and is given by*

$$\mathbb{E}(C(u)) = \sqrt{\frac{-2r''(0)}{\pi}}\,\varphi_1(u). \tag{8.28}$$

PROOF: see [43], p. 194. □

Theorem 65 *Let $\boldsymbol{x}(t) = (x_1(t), \ldots, x_n(t))$ be a with probability one continuously differentiable stationary normal vector process with derivative process $\boldsymbol{x}'(t) = (x_1'(t), \ldots, x_n'(t))$. Assume that $\boldsymbol{x}(t)$ is centered and standardized such that its autocorrelation functions have the form as in equation (8.17), p. 124. Further is given a function $g(\boldsymbol{x}) = 1 - \alpha^T\boldsymbol{x}$ with $\sum_{i=1}^n \alpha_i^2 = 1$. Then we have for the expected value $\mathbb{E}(C(\beta G))$*

$$\mathbb{E}(C(\beta G)) = \varphi_1(\beta)\sqrt{\frac{2}{\pi}\cdot\sum_{i=1}^n \tilde{\sigma}_i^2\alpha_i^2}. \tag{8.29}$$

PROOF: The crossings of the process $\boldsymbol{x}(t)$ through the hypersurface $G(\beta) = \{\boldsymbol{x}; \beta - \sum_{i=1}^n \alpha_i x_i = 0\}$ correspond to the crossings of the one-dimensional process $y(t) = \alpha^T\boldsymbol{x}(t)$ through the level β. This process $y(t)$ is a stationary normal process with

$$\mathrm{var}(y(t)) = 1 \text{ and } \mathrm{var}(y'(t)) = \sum_{i=1}^n \alpha_i^2\tilde{\sigma}_i^2 \tag{8.30}$$

The assertion follows from theorem 64, p. 128. □

For the case that the surface through which the process moves consists of a number of hyperplanes in [74] an asymptotic approximation is derived.

Lemma 66 *Given is a continuously differentiable stationary normal vector process $\boldsymbol{x}(t) = (x_1(t), \ldots, x_n(t))$ with derivative process $\boldsymbol{x}'(t) = (x_1'(t), \ldots, x_n'(t))$. The process is in a standardized and centered form. $\mathbf{c}$ is an arbitrary n-dimensional vector. Then we have for the conditional distribution of $\mathbf{c}^T \boldsymbol{x}'(t)$ that*

1. $\mathbb{E}(\mathbf{c}^T \boldsymbol{x}'(t); \boldsymbol{x}(t) = \boldsymbol{x}) = -\mathbf{c}^T \boldsymbol{R}'(0)\boldsymbol{x}$,
2. $\mathrm{var}(\mathbf{c}^T \boldsymbol{x}'(t); \boldsymbol{x}(t) = \boldsymbol{x}) = \mathbf{c}^T(-\boldsymbol{R}''(0) - \boldsymbol{R}'(0)^T \boldsymbol{R}'(0))\mathbf{c}$.

PROOF: From theorem 24, p. 30 we get the result with equations 8.13, p. 123. □

Theorem 67 *Given is a with probability one continuously differentiable stationary normal vector process $\boldsymbol{x}(t) = (x_1(t), \ldots, x_n(t))$ with derivative process $\boldsymbol{x}'(t) = (x_1'(t), \ldots, x_n'(t))$. The process is standardized and centered. Further is given a twice continuously differentiable function $g : \mathbb{R}^n \to \mathbb{R}$ with the following properties:*

1. *$g(\mathbf{o}) > 0$ and $\min_{\boldsymbol{x} \in G} |\boldsymbol{x}| = 1$,*
2. *G is a compact $(n-1)$-dimensional C^2-manifold with $\nabla g(\boldsymbol{x}) \neq \mathbf{o}$ for all $\boldsymbol{x} \in G$ and G is oriented by the normal vector field $\boldsymbol{n}(\boldsymbol{x}) = |\nabla g(\boldsymbol{x})|^{-1} \nabla g(\boldsymbol{x})$,*
3. *There are exactly k points $\boldsymbol{x}^1, \ldots, \boldsymbol{x}^k$ on the manifold G with*

$$|\boldsymbol{x}^1| = \ldots = |\boldsymbol{x}^k| = \min_{\boldsymbol{x} \in G} |\boldsymbol{x}| = 1, \tag{8.31}$$

   and at all these points $\boldsymbol{x}^1, \ldots, \boldsymbol{x}^k$ all main curvatures $\kappa_{i,j}$ of the surface G are less than one.

4. *The expected number of crossings of $\boldsymbol{x}(t)$ through the hypersurface G during $(0,1)$ is given by the integral in (8.27), p. 127.*

Under these conditions for the expected number of crossings of the process $\boldsymbol{x}(t)$ through the manifolds βG during the time interval $(0,1)$ the following asymptotic relation is valid

$$\mathbb{E}(C(\beta G)) \sim \varphi_1(\beta)\sqrt{\frac{2}{\pi}} \sum_{i=1}^{k} \sqrt{\frac{\sigma_1^2(\boldsymbol{x}_i) + \sigma_2^2(\boldsymbol{x}_i)}{d_i}}, \ \beta \to \infty. \tag{8.32}$$

with

$$d_i = \prod_{j=1}^{n-1}(1 - \kappa_{i,j}), \tag{8.33}$$

$$\sigma_1^2(\boldsymbol{x}) = \boldsymbol{n}^T(\boldsymbol{x})(-\boldsymbol{R}''(0) - \boldsymbol{R}'(0)^T \boldsymbol{R}'(0))\boldsymbol{n}(\boldsymbol{x}), \tag{8.34}$$

$$\sigma_2^2(\boldsymbol{x}) = \boldsymbol{n}^T(\boldsymbol{x})\boldsymbol{R}'(0)^T(\boldsymbol{I}_n + \boldsymbol{G}(\boldsymbol{x}))\boldsymbol{R}'(0)\boldsymbol{n}(\boldsymbol{x}), \tag{8.35}$$

$$\boldsymbol{G}(\boldsymbol{x}) = \left(|\nabla g(\boldsymbol{x})|^{-1} g^{ij}(\boldsymbol{x})\right)_{i,j=1,\ldots,n}. \tag{8.36}$$

PROOF: A proof of this theorem using theorem 42, p. 59 is given in [24]. Another proof which uses theorem 44, p. 62 is given in [25]. Here we give a short outline of a proof.

We consider the integral

$$J(\beta) = \int_{\beta G} \mathbb{E}(|\boldsymbol{n}^T(\boldsymbol{y})\boldsymbol{y}'(t)|; \boldsymbol{y}(t) = \boldsymbol{y})\varphi_n(\boldsymbol{y})ds_{\beta G}(\boldsymbol{y}). \tag{8.37}$$

Define now

$$\mu(\boldsymbol{x}) = \mathbb{E}(\boldsymbol{n}^T(\boldsymbol{x})\boldsymbol{x}'(t); \boldsymbol{x}(t) = \boldsymbol{x}) = -\boldsymbol{n}^T(\boldsymbol{x})\boldsymbol{R}'(0)\boldsymbol{x} \tag{8.38}$$

$$\sigma_1^2(\boldsymbol{x}) = \mathrm{var}(\boldsymbol{n}^T(\boldsymbol{x})\boldsymbol{x}'(t); \boldsymbol{x}(t) = \boldsymbol{x}) = \boldsymbol{n}^T(\boldsymbol{x})(-\boldsymbol{R}''(0) - \boldsymbol{R}'(0)^T\boldsymbol{R}'(0))\boldsymbol{n}(\boldsymbol{x}).$$

We write the conditional expectation in the form as integral in the form

$$J(\beta) = \int_{\beta G}\int_{\mathbb{R}} \frac{|w|}{\sigma_1(\boldsymbol{y})}\varphi_1((w - \mu(\boldsymbol{y}))/\sigma_1(\boldsymbol{y}))\varphi_n(\boldsymbol{y})dw ds_{\beta G}(\boldsymbol{y}). \tag{8.39}$$

Substituting $w \mapsto z = \beta^{-1}(w - \mu(\boldsymbol{y}))/\sigma_1(\boldsymbol{y})$ and $\boldsymbol{y} \mapsto \boldsymbol{x} = \beta^{-1}\boldsymbol{y}$ yields

$$= \beta^{-(n+1)} \int_G \int_{\mathbb{R}} |\sigma_1(\boldsymbol{x})z + \mu(\boldsymbol{x})|\varphi_1(\beta z)\varphi_n(\beta \boldsymbol{x})dz ds_G(\boldsymbol{x}).$$

Here we use lemma 66, p. 128 which gives that $\mu(\beta\boldsymbol{x}) = \beta\mu(\boldsymbol{x})$ and $\sigma_1(\beta\boldsymbol{x}) = \sigma(\boldsymbol{x})$.

Define now in $\mathbb{R}^{n+1}$ the n-dimensional C^2-manifold $\tilde{G}$ by $\tilde{G} = \{(\boldsymbol{x}, z) \in \mathbb{R}^{n+1}; g(\boldsymbol{x}) = 0\}$ so the last integral can be written as a surface integral over this surface

$$\begin{aligned} J(\beta) &= \beta^{-(n+1)} \int_{\tilde{G}} |\sigma_1(\boldsymbol{x})z + \mu(\boldsymbol{x})|\varphi_1(\beta z)\varphi_n(\beta\boldsymbol{x})ds_{\tilde{G}}(\boldsymbol{x}, z) \qquad (8.40) \\ &= \beta^{-(n+1)}(2\pi)^{-(n+1)/2} \int_{\tilde{G}} |\sigma_1(\boldsymbol{x})z + \mu(\boldsymbol{x})| \exp(-\frac{\beta^2}{2}(|\boldsymbol{x}|^2 + z^2))ds_{\tilde{G}}(\boldsymbol{x}, z). \end{aligned}$$

Define

$$f(\boldsymbol{x}, z) = -\frac{1}{2}(|\boldsymbol{x}|kz^2), \tag{8.41}$$

$$g_1(\boldsymbol{x}, z) = \sigma_1(\boldsymbol{x})z + \mu(\boldsymbol{x}), \tag{8.42}$$

$$\boldsymbol{H}(\boldsymbol{x}) = \boldsymbol{I}_n + \boldsymbol{G}(\boldsymbol{x}). \tag{8.43}$$

The function $f(\boldsymbol{x}, z)$ has on this surface exactly at the points $(\boldsymbol{x}^1, 0), \ldots, (\boldsymbol{x}^k, 0)$ global maxima. Since at $\boldsymbol{x}^l$ the function $|\boldsymbol{x}|^2$ has a minimum with respect to G, we have

$$\nabla|\boldsymbol{x}^l|^2 = 2\boldsymbol{x}^l = -2\boldsymbol{n}(\boldsymbol{x}^l). \tag{8.44}$$

From this follows, since $\boldsymbol{R}'(0)$ is a skew-symmetric matrix that $\mu(\boldsymbol{x}^l) = \boldsymbol{n}(\boldsymbol{x}^l)\boldsymbol{R}'(0)\boldsymbol{x}^l = 0$. The function $g_1(\boldsymbol{x}, z)$ has therefore the value 0 at these maximum points. Therefore we can use theorem 52, p. 80 to calculate the integral. For this we need the first derivatives of the function $\mu(\boldsymbol{x})$ and the gradient $\nabla g_1(\boldsymbol{x}, z)$. We get

$$\nabla g_1(\boldsymbol{x}, z) = (\nabla \sigma_1(\boldsymbol{x})z + \nabla \mu(\boldsymbol{x}), \sigma_1(\boldsymbol{x})). \tag{8.45}$$

We have

$$\mu(\boldsymbol{x}) = -|\nabla g(\boldsymbol{x})|^{-1} \sum_{i,j=1}^{n} g^i(\boldsymbol{x}) r'_{ij}(0) x_j. \tag{8.46}$$

Differentiating with respect to x_k gives

$$\frac{\partial \mu(\boldsymbol{x})}{\partial x_k} = -\left(\frac{\partial |\nabla g(\boldsymbol{x})|^{-1}}{\partial x_k} \sum_{i,j=1}^{n} g^i(\boldsymbol{x}) r'_{ij}(0) x_j + \right. \tag{8.47}$$

$$\left. + |\nabla g(\boldsymbol{x})|^{-1} \sum_{i,j=1}^{n} g^{ik}(\boldsymbol{x}) r'_{ij}(0) x_j + |\nabla g(\boldsymbol{x})|^{-1} \sum_{i,j=1}^{n} g^i(\boldsymbol{x}) r'_{ij}(0) \delta_{jk} \right). \tag{8.48}$$

Now we use equation (8.44), p. 130 and replace $-\boldsymbol{x}^l$ by $\boldsymbol{n}(\boldsymbol{x}^l)$. This gives

$$\frac{\partial \mu(\boldsymbol{x})}{\partial x_k} = -\left(\frac{\partial |\nabla g(\boldsymbol{x})|^{-1}}{\partial x_k} \sum_{i,j=1}^{n} g^i(\boldsymbol{x}) r'_{ij}(0) g^j(\boldsymbol{x}) |\nabla g(\boldsymbol{x})|^{-1} \right.$$

$$\left. + |\nabla g(\boldsymbol{x})|^{-1} \sum_{i,j=1}^{n} g^{ik}(\boldsymbol{x}) r'_{ij}(0) g^j(\boldsymbol{x}^l) |\nabla g(\boldsymbol{x}^l)|^{-1} + |\nabla g(\boldsymbol{x})|^{-1} \sum_{i,j=1}^{n} g^i(\boldsymbol{x}) r'_{ij}(0) \delta_{jk} \right).$$

Since $\boldsymbol{R}'(0)$ is a skew-symmetric matrix the first summand vanishes

$$\nabla \mu(\boldsymbol{x}^l) = -\left(-|\nabla g(\boldsymbol{x})|^{-1} \sum_{i,j=1}^{n} g^{ik}(\boldsymbol{x}) r'_{ij}(0) g^j(\boldsymbol{x}^l) |\nabla g(\boldsymbol{x}^l)|^{-1} \right.$$

$$\left. + |\nabla g(\boldsymbol{x})|^{-1} \sum_{i,j=1}^{n} g^i(\boldsymbol{x}) r'_{ij}(0) \delta_{jk} \right).$$

Writing this in matrix form gives

$$\nabla \mu(\boldsymbol{x}^l) = (\boldsymbol{I}_n + \boldsymbol{G}(\boldsymbol{x}^l)) \boldsymbol{R}'(0) \boldsymbol{n}(\boldsymbol{x}^l) = \boldsymbol{H}(\boldsymbol{x}^l) \boldsymbol{R}'(0) \boldsymbol{n}(\boldsymbol{x}^l) \tag{8.49}$$

and so we obtain

$$\nabla g_1(\boldsymbol{x}^l, 0) = (\boldsymbol{H}(\boldsymbol{x}^l) \boldsymbol{R}'(0) \boldsymbol{n}(\boldsymbol{x}^l), \sigma_1(\boldsymbol{x}^l)). \tag{8.50}$$

The generalized inverse of the Hessian $\boldsymbol{H}_{\tilde{f}}(\boldsymbol{x},z)$ of the function $\tilde{f} = f(\boldsymbol{x},z) - |\nabla g(\boldsymbol{x}^l)|^{-1} g(\boldsymbol{x})$ at $(\boldsymbol{x}^l, 0)$ is given by

$$\boldsymbol{H}_{\tilde{f}}^{-}(\boldsymbol{x},z) = \begin{pmatrix} & & 0 \\ -\boldsymbol{H}^{-}(\boldsymbol{x}^l) & & \vdots \\ & & 0 \\ & & 0 \\ 0\ldots 0 & & -1 \end{pmatrix}. \tag{8.51}$$

This gives for the determinant $\det(\boldsymbol{H}^*(\boldsymbol{x}^l))$

$$|\det(\boldsymbol{H}^*(\boldsymbol{x}^l))| = |\prod_{j=1}^{n-1}(1-\kappa_{i,j})| = d_i. \tag{8.52}$$

Using theorem 52, p. 80 gives the result. We obtain for the crossing rate

$$\mathbb{E}(C(\beta G)) \sim \sqrt{\frac{2}{\pi}}\varphi_1(-\beta)\sum_{i=1}^{k}\left|\frac{\nabla g_1(\boldsymbol{x}^l,0)^T \boldsymbol{H}_{\tilde{f}}^{-}(\boldsymbol{x}^l,0)\nabla g_1(\boldsymbol{x}^l,0)}{d_i}\right|^{1/2}$$

$$= \sqrt{\frac{2}{\pi}}\varphi_1(-\beta)\sum_{i=1}^{k}\left|\frac{\boldsymbol{n}^T(\boldsymbol{x}^l)\boldsymbol{R}'(0)^T\boldsymbol{H}(\boldsymbol{x}^l)\boldsymbol{H}(\boldsymbol{x}^l)^{-}\boldsymbol{H}(\boldsymbol{x}^l)\boldsymbol{R}'(0)\boldsymbol{n}(\boldsymbol{x}^l)+\sigma_1^2(\boldsymbol{x}^l)}{d_i}\right|^{1/2}$$

From the property $\boldsymbol{A}\boldsymbol{A}^{-}\boldsymbol{A} = \boldsymbol{A}$ follows then

$$= \sqrt{\frac{2}{\pi}}\varphi_1(-\beta)\sum_{i=1}^{k}\left|\frac{|(\boldsymbol{n}^T(\boldsymbol{x}^l)\boldsymbol{R}'(0)^T\boldsymbol{H}(\boldsymbol{x}^l)\boldsymbol{R}'(0)\boldsymbol{n}(\boldsymbol{x}^l)+\sigma_1^2(\boldsymbol{x}^l)|}{d_i}\right|^{1/2} \tag{8.53}$$

$$= \sqrt{\frac{2}{\pi}}\varphi_1(-\beta)\sum_{i=1}^{k}\sqrt{\frac{\sigma_2^1(\boldsymbol{x}^l)+\sigma_1^2(\boldsymbol{x}^l)}{d_i}}.$$

□

Here $\sigma_1^2(\boldsymbol{x}^l)$ is the conditonal variance of $\boldsymbol{n}^T(\boldsymbol{x})\boldsymbol{x}'(t)$ under the condition $\boldsymbol{x}(t) = \boldsymbol{x}^l$. $\sigma_2^2(\boldsymbol{x}^l)$ is the variance of the mean $\mu(\boldsymbol{x})$ if $\boldsymbol{x}$ is near $\boldsymbol{x}^l$. The variance of $\boldsymbol{n}^T(\boldsymbol{x})\boldsymbol{x}'(t)$ is decomposed into two components.

The following theorem was proved by Buss [38] for the case of independent component processes and constant curvatures. The generalization is new.

Theorem 68 *Given is a twice continuously differentiable function $g : \mathbb{R}^n \to \mathbb{R}$ with $g(\mathbf{o}) > 0$. The set M of all points $\boldsymbol{x} \in G = \{\boldsymbol{x}; g(\boldsymbol{x}) = 0\}$ with $|\boldsymbol{x}| = \min_{\boldsymbol{y} \in G} |\boldsymbol{y}| = 1$ is a compact k-dimensional submanifold of the $(n-1)$-dimensional manifold G with nonvanishing gradient $\nabla g(\boldsymbol{x})$, which is oriented by $\boldsymbol{n}(\boldsymbol{x}) = |\nabla g(\boldsymbol{x})|^{-1} \nabla g(\boldsymbol{x})$. Assume further that:*

1. *The $(n-k-1) \times (n-k-1)$-matrix $\boldsymbol{H}^*(\boldsymbol{x})$ is regular for all $\boldsymbol{x} \in M$ It is defined by:*
$$\boldsymbol{H}^*(\boldsymbol{x}) = \boldsymbol{A}^T(\boldsymbol{x})\boldsymbol{H}(\boldsymbol{x})\boldsymbol{A}(\boldsymbol{x})$$
with
$$\boldsymbol{H}(\boldsymbol{x}) = (-\delta_{ij} - |\nabla g(\boldsymbol{x})|^{-1} \cdot g^{ij}(\boldsymbol{x}))_{i,j=1,\dots,n}$$
and the $n \times (n-k-1)$-matrix $\boldsymbol{A}(\boldsymbol{x}) = (\boldsymbol{a}_1(\boldsymbol{x}), \dots, \boldsymbol{a}_{n-k-1}(\boldsymbol{x}))$ is composed of $n-k-1$ vectors being an orthonormal basis of the subspace of $T_G(\boldsymbol{x})$ which is orthogonal to $T_M(\boldsymbol{x})$.

2. *The expected number of crossings of $\boldsymbol{x}(t)$ through the hypersurface G during $(0,1)$ is given by the integral in (8.27), p. 127.*

Then the expected number of crossings of $\boldsymbol{x}(t)$ through βG during $(0,1)$ has the following asymptotic approximation

$$\mathbb{E}(C(\beta G)) \sim \chi_{k+1}(\beta) \cdot \frac{\Gamma((k+1)/2)}{2\pi^{(k+1)/2}} \cdot \int_M \sqrt{\frac{\sigma_1^2(\boldsymbol{x}) + \sigma_2^2(\boldsymbol{x})}{|\det(\boldsymbol{H}^*(\boldsymbol{x}))|}} ds_M(\boldsymbol{x}) \quad (8.54)$$

with

1. $\sigma_1^2(\boldsymbol{x}) = \boldsymbol{n}^T(\boldsymbol{x})(-\boldsymbol{R}''(0) - \boldsymbol{R}'(0)^T \boldsymbol{R}'(0))\boldsymbol{n}(\boldsymbol{x})$,
2. $\sigma_2^2(\boldsymbol{x}) = \boldsymbol{n}^T(\boldsymbol{x})\boldsymbol{R}'(0)^T(\boldsymbol{I}_n + \boldsymbol{G}(\boldsymbol{x}))\boldsymbol{R}'(0)\boldsymbol{n}(\boldsymbol{x})$

PROOF: Details are given in [31]. □

8.4.1 Examples

EXAMPLE: Given is a continuously differentiable two dimensional normal process $\boldsymbol{x}(t) = (x_1(t), x_2(t))$ with independent components. The variances of the process and the derivative process $\boldsymbol{x}'(t) = (x_1'(t), x_2'(t))$ are given by

$$\text{var}(x_i(t)) = \text{var}(x_i'(t)) = 1 \text{ for } i = 1, 2.$$

The limit state function is

$$g(x_1, x_2) = 1 - (\frac{x_1^2}{a^2} + x_2^2) \text{ with } 0 \le a \le 1.$$

If $a = 0$ the crossings through βG are exactly the crossings of the process $x_2(t)$ across the level β.

If $0 < a < 1$ there are exactly two points on the ellipse G where the probability density of the process is maximal, the points $(0,1)$ and $(0,-1)$. We get as asymptotic approximation for the crossing rate

$$\mathbb{E}(C(\beta G)) \sim 2\varphi_1(\beta)\sqrt{\frac{2}{\pi}}\sqrt{\frac{1}{1-a^{-2}}}.$$

If $a = 1$ the density is maximal on the whole circle. This gives for the crossing rate

$$\mathbb{E}(C(\beta G)) \sim \chi_2(\beta)\sqrt{\frac{2}{\pi}}.$$

□

EXAMPLE: Given is an n-dimensional continuouly differentiable normal process $\boldsymbol{x}(t)$. Further is given a function $g(\boldsymbol{x}) = 1 - \sum_{i=1}^{n} \gamma_i x_i^2$ with $0 < \gamma_1 \le \ldots \le \gamma_{n-1} < \gamma_n = 1$. Then there are exactly two points $\boldsymbol{y}^1 = (0,\ldots,0,1)$ and $\boldsymbol{y}^2 = -\boldsymbol{y}^1$ on the surface $G = \{\boldsymbol{x}; g(\boldsymbol{x}) = 0\}$ with minimal distance to the origin.

We get for $k = 1, 2$:

1. $d_k = \prod_{i=1}^{n-1}(1-\gamma_i)$,
2. $\sigma_1^2(\boldsymbol{y}^k) = -r''_{nn}(0) - \sum_{i=1}^{n-1} r'^2_{in}(0)$,
3. $\sigma_2^2(\boldsymbol{y}^k) = \sum_{i=1}^{n-1}(1-\gamma_i) r'^2_{in}(0)$.

This gives

$$\mathbb{E}(C(\beta G)) \sim \varphi_1(\beta) 2\sqrt{\frac{2}{\pi}}\sqrt{\frac{-r''^2_{nn}(0) - \sum_{i=1}^{n-1}\gamma_i r'^2_{in}(0)}{\prod_{i=1}^{n-1}(1-\gamma_i)}}.$$

□

A further numerical example can be found in [21].

Bibliography

[1] M. Abramowitz and I.A. Stegun. *Handbook of Mathematical Functions.* Dover, New York, 1965.

[2] J.M.P. Albin. *On Extremal Theory for Nondifferentiable Stationary Processes.* PhD thesis, University of Lund, Department of Mathematical Statistics, Lund (Sweden), 1987.

[3] J.M.P. Albin. On extremal theory for stationary processes. *Annals of Probability*, 18:92–128, 1990.

[4] S.T. Ariaratnam, G.I. Schuëller, and I. Elishakoff, editors. *Stochastic Structural Dynamics.* Elsevier, London, 1988.

[5] G. Aumann and O. Haupt. *Einführung in die reelle Analysis*, volume II. De Gruyter, 1979.

[6] B.M. Ayyub and K.-L. Lai. Structural reliability assessment with ambiguity and vagueness in failure. *Naval Engineers Journal*, 104:21–35, 1992.

[7] O.E. Barndorff-Nielsen and D.R. Cox. *Asymptotic Techniques for Use in Statistics.* Chapman and Hall, London, 1989.

[8] Yu.K. Belyaev. On the number of exits across a boundary of a region by a vector process. *Theory of Probability and Applications*, 13:320–324, 1968.

[9] Y. Ben-Haim and I. Elishakoff. *Convex Models of Uncertainty in Applied Mechanics.* Elsevier, Amsterdam, 1990.

[10] J.R. Benjamin and C.A. Cornell. *Probability, Statistics and Decision for Civil Engineers.* McGraw Hill, New York, N.Y., 1970.

[11] L. Berg. *Asymptotische Darstellungen und Entwicklungen.* VEB Deutscher Verlag der Wissenschaften, Berlin (GDR), 1968.

[12] S. M. Berman. *Sojourns and Extremes of Stochastic Processes.* Wadsworth & Brooks/Cole Advanced Books & Software, Pacific Grove, CA, 1992.

[13] P. Billingsley. *Probability and Measure.* Wiley, New York, first edition, 1979.

[14] H. Birndt and W.-D. Richter. Vergleichende Betrachtungen zur Bestimmung des asymptotischen Verhaltens mehrdimensionaler Laplace-Gauß-Integrale. *Zeitschrift für Analysis und ihre Anwendungen*, 4(3):269–276, 1985.

[15] N. Bleistein and R.A. Handelsman. *Asymptotic Expansions of Integrals.* Dover Publications Inc., New York, 1986. Reprint of the edition by Holt, Rinehart and Winston, New York, 1975.

[16] V.V. Bolotin. *Statistical Methods in Structural Mechanics.* Holden-Day, San Francisco, 1969.

[17] V.V. Bolotin. *Wahrscheinlichkeitsmethoden zur Berechnung von Konstruktionen.* VEB-Verlag für das Bauwesen, Berlin (GDR), 1981.

[18] E. Bolthausen. Laplace approximations for sums of independent random vectors. *Probability Theory and Related Fields*, 72:305–318, 1986.

[19] E. Bolthausen. Laplace approximations for sums of independent random vectors. Part II. Degenerate maxima and manifolds. *Probability Theory and Related Fields*, 76:167–206, 1987.

[20] K. Breitung. An asymptotic formula for the failure probability. In *DIALOG 82-6*, pages 19–45, Lyngby (Denmark), 1982. Department of Civil Engineering, Danmarks Ingeniorakademi.

[21] K. Breitung. Asymptotic approximations for multinormal domain and surface integrals. In G. Augusti et al., editor, *Proceedings of the 4th International Conference on Applications of Statistics and Probability on Soil and Structural Engineering*, pages 755–767 (Vol.2), Firenze (Italy), 1983. Universita' di Firenze, Pitagora Editrice.

[22] K. Breitung. *Edgeworthentwicklungen für die Dichten von differenzierbaren gefilterten Poissonprozessen.* PhD thesis, Fakultät 10, Ludwig-Maximilians-Universität München, 1983.

[23] K. Breitung. Asymptotic approximations for multinormal integrals. *Journal of the Engineering Mechanics Division ASCE*, 110(3):357–366, 1984.

[24] K. Breitung. Asymptotic approximations for the crossing rates of stationary Gaussian vector processes. Technical Report 1984:1, Dept. of Math. Statistics, Lund (Sweden), 1984.

[25] K. Breitung. Asymptotic crossing rates for stationary Gaussian vector processes. *Stochastic Processes and Applications*, 29:195–207, 1988.

[26] K. Breitung. Asymptotic approximations for probability integrals. *Journal of Probabilistic Engineering Mechanics*, 4(4):187–190, 1989.

[27] K. Breitung. The extreme value distribution of non-stationary vector processes. In A. H.-S. Ang, M. Shinozuka, and G.I. Schuëller, editors, *Proceedings of ICOSSAR '89 5th Int'l Conf. on structural safety and reliability*, volume II, pages 1327–1332. American Society of Civil Engineers, 1990.

[28] K. Breitung. Parameter sensitivity of failure probabilities. In A. Der-Kiureghian and P. Thoft-Christensen, editors, *Reliability and Optimization of Structural Systems '90, Proceedings of the 3rd IFIP WG 7.5 Conference, Berkeley, California*, pages 43–51, New York, 1991. Springer. Lecture Notes in Engineering 61.

[29] K. Breitung. Probability approximations by log likelihood maximization. *Journal of the Engineering Mechanics Division ASCE*, 117(3):457–477, 1991.

[30] K. Breitung. A criticism of statistical methods in probabilistic models in structural reliability. In Y.K. Lin, editor, *Probabilistic Mechanics and Structural and Geotechnical Reliability, Proceedings of the sixth specialty conference*, pages 236–239. American Society of Civil Engineers, 1992.

[31] K. Breitung. Crossing rates of Gaussian vector processes. In *Transactions of the 11th Prague Conference on Information Theory, Statistical Decision Functions and Random Processes 1990*, volume I, pages 303–314, Prague, Czech. Rep., 1992. Academia.

[32] K. Breitung and L. Faravelli. Log-likelihood maximization and response surface reliability assessment. *Nonlinear Dynamics*, 5(3):273–286, 1994.

[33] K. Breitung and M. Hohenbichler. Asymptotic approximations for multivariate integrals with an application to multinormal probabilities. *Journal of Multivariate Analysis*, 30:80–97, 1989.

[34] K. Breitung and Y. Ibrahim. Problems of statistical inference in structural reliability. In G.I. G.I. Schuëller, M. Shinozuka, and J.T.P. Yao, editors, *Proceedings of ICOSSAR '93 6th Int'l Conf. on structural safety and reliability*, volume II, pages 1223–1226, Rotterdam, 1994. A.A. Balkema.

[35] K. Breitung and W.-D. Richter. A geometric approach to an asymptotic expansion for large deviation probabilities of Gaussian random vectors, 1993. submitted to the Journal of Multivariate Analysis.

[36] C.G. Bucher. Adaptive sampling - an iterative fast Monte Carlo method. *Structural Safety*, 5:119–126, 1988.

[37] E.V. Bulinskaya. On the mean number of crossings of a level by a stationary Gaussian process. *Theory of Probability and its Applications*, 6:435–438, 1961.

[38] A. Buss. *Crossings of non-Gaussian processes with reliability applications.* PhD thesis, Department of Structural Engineering, Cornell University, Ithaca, N.Y., 1986.

[39] F. Casciati and L. Faravelli. *Fragility Analysis of Complex Structures.* Research Studies Press Ltd., Taunton, UK, 1991.

[40] E. Castillo. *Extreme Value Theory in Engineering.* Academic Press, Boston, 1989.

[41] R.-H. Cherng and Y.K. Wen. Reliability of uncertain nonlinear trusses under random excitation. II. *Journal of the Engineering Mechanics Division ASCE*, 120(4):748–757, 1994.

[42] C.A. Cornell. A probability-based structural code. *Journal of the American Concrete Institute*, 66(12):974–985, 1969.

[43] H. Cramer and M.R. Leadbetter. *Stationary and Related Stochastic Processes.* Wiley, New York, 1967.

[44] A. Der Kiureghian. Measures of structural safety under imperfect states of knowledge. *Journal of the Engineering Mechanics Division ASCE*, 115(5):1119–1140, 1989.

[45] A. Der Kiureghian and M. De Stefano. Efficient algorithm for second-order reliability analysis. *Journal of the Engineering Mechanics Division ASCE*, 117(12):2904–2923, 1991.

[46] A. Der Kiureghian, H.Z. Lin, and S.-J. Hwang. Second-order reliability approximations. *Journal of the Engineering Mechanics Division ASCE*, 113(8):1208–1225, 1987.

[47] O. Ditlevsen. Structural reliability and the invariance problem. Technical Report Research Report No. 22, Solid Mechanics Division, University of Waterloo, Waterloo, Canada, 1973.

[48] O. Ditlevsen. Generalized second-moment reliability index. *Journal of the Engineering Mechanics Division; ASCE*, 107(6):1191–1209, 1979.

[49] O. Ditlevsen. Narrow reliability bounds for structural systems. *Journal of Structural Mechanics*, 7:435–451, 1979.

[50] O. Ditlevsen. Principle of normal tail approximation. *Journal of the Engineering Mechanics Division ASCE*, 107(6):1191–1209, 1981.

[51] O. Ditlevsen. Taylor expansion of series system reliability. *Journal of the Engineering Mechanics Division ASCE*, 108:293–307, 1984.

[52] K. Dolinski. First-order second-moment approximation in reliability of systems: critical review and alternative approach. *Structural Safety*, 1:211–213, 1983.

[53] R.S. Ellis and J.S. Rosen. Asymptotic analysis of Gaussian integrals. I. Isolated minimum points. *Transactions of the American Mathematical Society*, 273(2):447–481, 1982.

[54] R.S. Ellis and J.S. Rosen. Asymptotic analysis of Gaussian integrals, II: Manifold of minimum points. *Zeitschrift für Wahrscheinlichkeitstheorie und verwandte Gebiete*, 273(2):156–181, 1982.

[55] R.S. Ellis and J.S. Rosen. Laplace method for Gaussian integrals with an application to statistical mechanics. *The Annals of Probability*, 10(1):47–66, 1982.

[56] S. Engelund and R. Rackwitz. A benchmark study on importance sampling techniques in structural reliability. *Structural Safety*, 12:255–276, 1993.

[57] A. Erdélyi. *Asymptotic Expansions.* Dover, New York, 1956.

[58] L. Faravelli. Response-surface approach for reliability analysis. *Journal of the Engineering Mechanics Division ASCE*, 115(12):2763–2781, 1989.

[59] M.V. Fedoryuk. *Metod perevala (The saddlepoint method).* Nauka, Moscow, USSR, 1977. In Russian.

[60] R. Ferrando, M. Dogliani, and R. Cazzulo. Maxima of mooring forces: an approach based on outcrossing methods. Technical Report RR 224, Registro italiano navale, Genova (Italy), 1989.

[61] B. Fiessler, H.-J. Neumann, and R. Rackwitz. Quadratic limit states in structural reliability. *Journal of the Engineering Mechanics Division ASCE*, 105(4):661–676, 1979.

[62] W. Fleming. *Functions of Several Variables.* Springer, New York, 1977.

[63] J. Focke. Asymptotische Entwicklungen mittels der Methode der stationären Phase. Berichte über die Verhandlungen der Sächsischen Akademie der Wissenschaften zu Leipzig, Mathematisch-naturwissenschaftliche Klasse, 101,3, 1953.

[64] A.M. Freudenthal. The safety of structures. *Transactions of the ASCE*, 112:1337–1397, 1947.

[65] A.M. Freudenthal. Safety and the probability of structural failure. *Transactions of the ASCE*, 121:1337–1397, 1956.

[66] W. Fulks and J.O. Sather. Asymptotics II. Laplace's method for multiple integrals. *Pacific Journal of Mathematics*, 11:185–192, 1961.

[67] F.A. Graybill. *Theory and Application of the Linear Model.* Wadsworth and Brooks/Cole, Pacific Grove, California, 1976.

[68] E.J. Gumbel. *Statistics of Extremes.* Columbia University Press, New York, 1958.

[69] A.M. Hasofer and N.C. Lind. An exact and invariant first-order reliability format. *Journal of the Engineering Mechanics Division ASCE*, 100(1):111–121, 1974.

[70] M.R. Hestenes. *Optimization Theory, The finite dimensional case*. Wiley, 1975.

[71] M. Hohenbichler, S. Gollwitzer, W. Kruse, and R. Rackwitz. New light on first-and second-order reliability methods. *Structural Safety*, 4:267–284, 1987.

[72] M. Hohenbichler and R. Rackwitz. Non-normal dependent vectors in structural safety. *Journal of the Engineering Mechanics Division ASCE*, 107(6):1227–1241, 1981.

[73] M. Hohenbichler and R. Rackwitz. First-order concepts in system reliability. *Structural Safety*, 1:177–188, 1983.

[74] M. Hohenbichler and R. Rackwitz. Asymptotic crossing rate of Gaussian vector processes into intersections of failure domains. *Probabilistic Engineering Mechanics*, 1(3):177–179, 1986.

[75] M. Hohenbichler and R. Rackwitz. Improvement of second-order reliability estimates by importance sampling. *Journal of the Engineering Mechanics Division ASCE*, 114(12):2195–2198, 1988.

[76] L.C. Hsu. On the asymptotic evaluation of a class of multiple integrals involving a parameter. *Duke Mathematical Journal*, 15:625–634, 1948.

[77] Y. Ibrahim. Observations on applications of importance sampling in structural reliability analysis. *Structural Safety*, 9(4):269–282, 1991.

[78] A.I. Johnson. *Strength, Safety and Economical Dimensions of Structures*. Statens Kommitté för Byggnadsforskning, Stockholm, Sweden, 1953. Meddelanden No. 22.

[79] D.S. Jones. *Generalized Functions*. McGraw-Hill, London, 1966.

[80] A. Karamchandani and C.A. Cornell. Adaptive hybrid conditional expectation approaches for reliability estimation. *Structural Safety*, 11(1):59–74, 1991.

[81] M.R. Leadbetter. On crossings of levels and curves by a wide class of stochastic processes. *Annals of Mathematical Statistics*, 37:260–267, 1966.

[82] M.R. Leadbetter, G. Lindgren, and H. Rootzén. *Extremes and Related Properties of Random Sequences and Processes*. Springer, New York, 1983.

[83] T. Levi-Civita. *The Absolute Differential Calculus*. Dover, New York, 1977.

[84] Y.K. Lin. *Probabilistic theory of structural dynamics*. R. Krieger Publishing Co., Malabar, Florida, 1976.

[85] G. Lindgren. Extremal ranks and transformation of variables for extremes of functions of multivariate Gaussian processes. *Stochastic Processes and Applications*, 17:285–312, 1984.

[86] D.V. Lindley. The use of prior probability distributions in statistical inference and decisions. In *Proceedings of the 4th Berkeley Symposium*, volume 1, pages 453–468, 1961.

[87] P.-L. Liu and A. Der Kiureghian. Optimization algorithms for structural reliability. *Structural Safety*, 9(3):161–177, 1991.

[88] H. Madsen. Omission sensitivity factors. *Structural Safety*, 5:35–45, 1988.

[89] H. Madsen, S. Krenk, and N.C. Lind. *Methods of Structural Safety*. Prentice-Hall Inc., Englewood Cliffs, N.J., 1986.

[90] M.A. Maes. Codification of design load criteria subject to modeling uncertainty. *Journal of the Structural Engineering Division, ASCE*, 117(10):2988–3007, 1991.

[91] M.A. Maes, K. Breitung, and D.J. Dupuis. Asymptotic importance sampling. *Structural Safety*, 12:167–186, 1993.

[92] M.A. Maes and D.J. Dupuis. Computational aspects of civil engineering reliability analysis. In *Proceedings of the 1991 Annual Conference, Canadian Society of Civil Engineers, Vancouver*, volume 2, pages 221–230, 1991.

[93] J.R. Magnus and H. Neudecker. *Matrix Differential Calculus with Applications in Statistics and Econometrics*. Wiley, New York, 1988.

[94] M.B. Marcus. Level crossings of a stochastic process with absolutely continous sample paths. *Annals of Probability*, 5:52–71, 1977.

[95] K. Marti. Approximations and derivatives of probabilities in structural design. *Zeitschrift für angewandte Mathematik und Mechanik*, 72(6):T575–T578, 1992.

[96] K. Marti. Stochastic optimization in structural design. *Zeitschrift für angewandte Mathematik und Mechanik*, 72(6):T452–T464, 1992.

[97] G. Matheron. *Estimating and Choosing*. Springer, Berlin, 1989.

[98] M. Mayer. *Die Sicherheit der Bauwerke*. Springer, Berlin, 1926.

[99] R.E. Melchers. *Structural Reliability, Analysis and Prediction*. Wiley, New York, 1987.

[100] K.S. Miller. *Multidimensional Gaussian Distributions*. Wiley, New York, 1964.

[101] K.S. Miller. *An Introduction to Vector Stochastic Processes*. Krieger, Huntington, N.Y., 1980.

[102] J.A. Munkres. *Analysis on Manifolds.* Addisson-Wesley, Redwood City, CA, 1991.

[103] D.C. Murdoch. *Linear Algebra for Undergraduates.* Wiley, New York, 1957.

[104] A. Naess. Bounding approximations to some quadratic limit states. *Journal of the Engineering Mechanics Division ASCE*, 113(10):1474–1492, 1987.

[105] F.W.J. Olver. *Asymptotics and Special Functions.* Academic Press, London, 1974.

[106] D.B. Owen. A table of normal integrals. *Communications in Statistics-Simulation and Computation*, B9(4):389–419, 1980.

[107] G. Pap and W.-D. Richter. Zum asymptotischen Verhalten der Verteilungen und der Dichten gewisser Funktionale Gauss'scher Zufallsvektoren. *Mathematische Nachrichten*, 135:119–124, 1988.

[108] A. Papoulis. *Probability, Random Variables and Stochastic Processes.* McGraw-Hill Kogakusha, Ltd., Tokyo, first edition, 1965.

[109] B.N. Pšeničnyj. Algorithms for general mathematical programming problems. *Cybernetics*, 6(5):120–125, 1970.

[110] R. Rackwitz and B. Fiessler. Structural reliability under combined random load sequences. *Computers and Structures*, 9:489–494, 1978.

[111] C.R. Rao. *Linear Statistical Inference and Its Applications.* Wiley, New York, second edition, 1973.

[112] C.R. Rao. *Lineare statistische Methoden und ihre Anwendungen.* Akademie-Verlag, Berlin, 1973.

[113] C.R. Rao and S.K. Mitra. *Generalized Inverse of Matrices and its Applications.* Wiley, New York, 1971.

[114] W.-D. Richter. Moderate deviations in special sets of $\mathbb{R}^k$. *Mathematische Nachrichten*, 113:339–354, 1983.

[115] W.-D. Richter. Laplace-Gauss integrals, Gaussian measure asymptotic behavior and probabilities of moderate deviations. *Zeitschrift für Analysis und ihre Anwendungen*, 4(3):257–267, 1985.

[116] W.-D. Richter. Remarks on moderate deviations in the multidimensional central limit theorem. *Mathematische Nachrichten*, 122:167–173, 1985.

[117] W.-D. Richter. Laplace integral and probabilities of moderate deviations in $\mathbb{R}^k$. In *Probability distributions and mathematical statistics*, pages 406–420, Tashkent, 1986. "Fan". In Russian.

[118] W.-D. Richter. Moderate deviations for a certain class of sets in finite dimensional space. *Theory of Probability and its Applications*, 31(1):174–175, 1986.

[119] W.-D. Richter. Zur Restgliedabschätzung im mehrdimensionalen integralen Zentralen Grenzwertsatz der Wahrscheinlichkeitstheorie. *Mathematische Nachrichten*, 135:103–117, 1988.

[120] M. Rosenblatt. Remarks on a multivariate transformation. *The Annals of Mathematical Statistics*, 23:470–472, 1952.

[121] E. Rosenblueth and L. Esteva. Reliability basis for some Mexican codes. *ACI Publication*, SP-31:1–41, 1972.

[122] H. Ruben. An asymptotic expansion for the multivariate normal distribution and Mill's ratio. *J. Res. Nat. Bur. Standards B*, 68(1):3–11, 1964.

[123] R.Y. Rubinstein. *Monte Carlo Optimization, Simulation and Sensitivity of Queuing Networks*. Wiley, New York, 1986.

[124] G.I. Schuëller. *Einführung in die Sicherheit und Zuverlässigkeit von Tragwerken*. Verlag Ernst u. Sohn, Munich, 1981.

[125] G.I. Schuëller and R. Stix. A critical appraisal of methods to determine failure probabilities. *Structural Safety*, 4:293–309, 1987.

[126] M. Shinozuka. Basic analysis of structural safety. *Journal of the Structural Division ASCE*, 109(3):721–740, 1983.

[127] L. Sirovich. *Techniques of Asymptotic Analysis*. Springer, New York, 1971.

[128] G. Spaethe. *Die Sicherheit tragender Baukonstruktionen*. Springer Verlag, Wien, New York, second edition, 1992.

[129] J. Stoer. *Einführung in die Numerische Mathematik I*. Springer, Berlin, 1972.

[130] J.A. Thorpe. *Elementary Topics in Differential Geometry*. Springer, New York, 1979.

[131] L. Tierney and J.B. Kadane. Accurate approximations for posterior moments and marginal densities. *Journal of the American Statistical Association*, 81:82–86, 1986.

[132] T.Y. Torng and Y.-T. Wu. Development of a probabilistic analysis methodology for structural reliability estimation. In *AIAA/ASME/ASCE/AHS/ASC 32nd Structures, Structural Dynamics, and Materials Conference*, pages 1271–1279. The American Institute of Aeronautics and Astronautics, 1991. part 2.

[133] L. Tvedt. Two second-order approximations to the failure probability. Technical Report Veritas Report RDIV/20-004-83, Det norkse Veritas, Oslo, Norway, 1983.

[134] L. Tvedt. Second-order reliability by an exact integral. In P. Thoft-Christensen, editor, *Reliability and Optimization of Structural Systems '88 Proceedings of the 2nd IFIP WG7.5 Conference London*, pages 377–384, New York, 1989. Springer, Lecture Notes in Engineering 48.

[135] L. Tvedt. Distribution of quadratic forms in normal space - application to structural reliability. *Journal of the Engineering Mechanics Division ASCE*, 116(6):1183–1198, 1990.

[136] S. Uryas'ev. A diffferentiation formula for integrals over sets given by inclusion. *Numerical Functional Analysis and Optimization*, 10:827–841, 1989.

[137] D. Veneziano, M. Grigoriu, and C.A. Cornell. Vector process models for system reliability. *Journal of the Engineering Mechanics Division ASCE*, 103(3):441–460, 1977.

[138] G.N. Watson. Harmonic functions associated with the parabolic cylinder. *Proceedings of the London Mathematical Society*, 17:116–148, 1918. Series 2.

[139] J.R. Westlake. *A Handbook of Numerical Matrix Inversion and Solution of Linear Equations.* Wiley, New York, 1968.

[140] R. Wong. *Asymptotic Approximations of Integrals.* Academic Press, San Diego, 1989.

[141] Y.-T. Wu and P. H. Wirsching. New algorithm for structural reliability estimation. *Journal of the Engineering Mechanics Division ASCE*, 113(9):1319–1336, 1987.

Index

Lecture Notes in Mathematics

For information about Vols. 1–1414
please contact your bookseller or Springer-Verlag

Vol. 1415: M. Langevin, M. Waldschmidt (Eds.), Cinquante Ans de Polynômes. Fifty Years of Polynomials. Proceedings, 1988. IX, 235 pages.1990.

Vol. 1416: C. Albert (Ed.), Géométrie Symplectique et Mécanique. Proceedings, 1988. V, 289 pages. 1990.

Vol. 1417: A.J. Sommese, A. Biancofiore, E.L. Livorni (Eds.), Algebraic Geometry. Proceedings, 1988. V, 320 pages. 1990.

Vol. 1418: M. Mimura (Ed.), Homotopy Theory and Related Topics. Proceedings, 1988. V, 241 pages. 1990.

Vol. 1419: P.S. Bullen, P.Y. Lee, J.L. Mawhin, P. Muldowney, W.F. Pfeffer (Eds.), New Integrals. Proceedings, 1988. V, 202 pages. 1990.

Vol. 1420: M. Galbiati, A. Tognoli (Eds.), Real Analytic Geometry. Proceedings, 1988. IV, 366 pages. 1990.

Vol. 1421: H.A. Biagioni, A Nonlinear Theory of Generalized Functions, XII, 214 pages. 1990.

Vol. 1422: V. Villani (Ed.), Complex Geometry and Analysis. Proceedings, 1988. V, 109 pages. 1990.

Vol. 1423: S.O. Kochman, Stable Homotopy Groups of Spheres: A Computer-Assisted Approach. VIII, 330 pages. 1990.

Vol. 1424: F.E. Burstall, J.H. Rawnsley, Twistor Theory for Riemannian Symmetric Spaces. III, 112 pages. 1990.

Vol. 1425: R.A. Piccinini (Ed.), Groups of Self-Equivalences and Related Topics. Proceedings, 1988. V, 214 pages. 1990.

Vol. 1426: J. Azéma, P.A. Meyer, M. Yor (Eds.), Séminaire de Probabilités XXIV, 1988/89. V, 490 pages. 1990.

Vol. 1427: A. Ancona, D. Geman, N. Ikeda, École d'Eté de Probabilités de Saint Flour XVIII, 1988. Ed.: P.L. Hennequin. VII, 330 pages. 1990.

Vol. 1428: K. Erdmann, Blocks of Tame Representation Type and Related Algebras. XV. 312 pages. 1990.

Vol. 1429: S. Homer, A. Nerode, R.A. Platek, G.E. Sacks, A. Scedrov, Logic and Computer Science. Seminar, 1988. Editor: P. Odifreddi. V, 162 pages. 1990.

Vol. 1430: W. Bruns, A. Simis (Eds.), Commutative Algebra. Proceedings. 1988. V, 160 pages. 1990.

Vol. 1431: J.G. Heywood, K. Masuda, R. Rautmann, V.A. Solonnikov (Eds.), The Navier-Stokes Equations – Theory and Numerical Methods. Proceedings, 1988. VII, 238 pages. 1990.

Vol. 1432: K. Ambos-Spies, G.H. Müller, G.E. Sacks (Eds.), Recursion Theory Week. Proceedings, 1989. VI, 393 pages. 1990.

Vol. 1433: S. Lang, W. Cherry, Topics in Nevanlinna Theory. II, 174 pages.1990.

Vol. 1434: K. Nagasaka, E. Fouvry (Eds.), Analytic Number Theory. Proceedings, 1988. VI, 218 pages. 1990.

Vol. 1435: St. Ruscheweyh, E.B. Saff, L.C. Salinas, R.S. Varga (Eds.), Computational Methods and Function Theory. Proceedings, 1989. VI, 211 pages. 1990.

Vol. 1436: S. Xambó-Descamps (Ed.), Enumerative Geometry. Proceedings, 1987. V, 303 pages. 1990.

Vol. 1437: H. Inassaridze (Ed.), K-theory and Homological Algebra. Seminar, 1987–88. V, 313 pages. 1990.

Vol. 1438: P.G. Lemarié (Ed.) Les Ondelettes en 1989. Seminar. IV, 212 pages. 1990.

Vol. 1439: E. Bujalance, J.J. Etayo, J.M. Gamboa, G. Gromadzki. Automorphism Groups of Compact Bordered Klein Surfaces: A Combinatorial Approach. XIII, 201 pages. 1990.

Vol. 1440: P. Latiolais (Ed.), Topology and Combinatorial Groups Theory. Seminar, 1985–1988. VI, 207 pages. 1990.

Vol. 1441: M. Coornaert, T. Delzant, A. Papadopoulos. Géométrie et théorie des groupes. X, 165 pages. 1990.

Vol. 1442: L. Accardi, M. von Waldenfels (Eds.), Quantum Probability and Applications V. Proceedings, 1988. VI, 413 pages. 1990.

Vol. 1443: K.H. Dovermann, R. Schultz, Equivariant Surgery Theories and Their Periodicity Properties. VI, 227 pages. 1990.

Vol. 1444: H. Korezlioglu, A.S. Ustunel (Eds.), Stochastic Analysis and Related Topics VI. Proceedings, 1988. V, 268 pages. 1990.

Vol. 1445: F. Schulz, Regularity Theory for Quasilinear Elliptic Systems and – Monge Ampère Equations in Two Dimensions. XV, 123 pages. 1990.

Vol. 1446: Methods of Nonconvex Analysis. Seminar, 1989. Editor: A. Cellina. V, 206 pages. 1990.

Vol. 1447: J.-G. Labesse, J. Schwermer (Eds), Cohomology of Arithmetic Groups and Automorphic Forms. Proceedings, 1989. V, 358 pages. 1990.

Vol. 1448: S.K. Jain, S.R. López-Permouth (Eds.), Non-Commutative Ring Theory. Proceedings, 1989. V, 166 pages. 1990.

Vol. 1449: W. Odyniec, G. Lewicki, Minimal Projections in Banach Spaces. VIII, 168 pages. 1990.

Vol. 1450: H. Fujita, T. Ikebe, S.T. Kuroda (Eds.), Functional-Analytic Methods for Partial Differential Equations. Proceedings, 1989. VII, 252 pages. 1990.

Vol. 1451: L. Alvarez-Gaumé, E. Arbarello, C. De Concini, N.J. Hitchin, Global Geometry and Mathematical Physics. Montecatini Terme 1988. Seminar. Editors: M. Francaviglia, F. Gherardelli. IX, 197 pages. 1990.

Vol. 1452: E. Hlawka, R.F. Tichy (Eds.), Number-Theoretic Analysis. Seminar, 1988–89. V, 220 pages. 1990.

Vol. 1453: Yu.G. Borisovich, Yu.E. Gliklikh (Eds.), Global Analysis – Studies and Applications IV. V, 320 pages. 1990.

Vol. 1454: F. Baldassari, S. Bosch, B. Dwork (Eds.), p-adic Analysis. Proceedings, 1989. V, 382 pages. 1990.

Vol. 1455: J.-P. Françoise, R. Roussarie (Eds.), Bifurcations of Planar Vector Fields. Proceedings, 1989. VI, 396 pages. 1990.

Vol. 1456: L.G. Kovács (Ed.), Groups – Canberra 1989. Proceedings. XII, 198 pages. 1990.

Vol. 1457: O. Axelsson, L.Yu. Kolotilina (Eds.), Preconditioned Conjugate Gradient Methods. Proceedings, 1989. V, 196 pages. 1990.

Vol. 1458: R. Schaaf, Global Solution Branches of Two Point Boundary Value Problems. XIX, 141 pages. 1990.

Vol. 1459: D. Tiba, Optimal Control of Nonsmooth Distributed Parameter Systems. VII, 159 pages. 1990.

Vol. 1460: G. Toscani, V. Boffi, S. Rionero (Eds.), Mathematical Aspects of Fluid Plasma Dynamics. Proceedings, 1988. V, 221 pages. 1991.

Vol. 1461: R. Gorenflo, S. Vessella, Abel Integral Equations. VII, 215 pages. 1991.

Vol. 1462: D. Mond, J. Montaldi (Eds.), Singularity Theory and its Applications. Warwick 1989, Part I. VIII, 405 pages. 1991.

Vol. 1463: R. Roberts, I. Stewart (Eds.), Singularity Theory and its Applications. Warwick 1989, Part II. VIII, 322 pages. 1991.

Vol. 1464: D. L. Burkholder, E. Pardoux, A. Sznitman, Ecole d'Eté de Probabilités de Saint- Flour XIX-1989. Editor: P. L. Hennequin. VI, 256 pages. 1991.

Vol. 1465: G. David, Wavelets and Singular Integrals on Curves and Surfaces. X, 107 pages. 1991.

Vol. 1466: W. Banaszczyk, Additive Subgroups of Topological Vector Spaces. VII, 178 pages. 1991.

Vol. 1467: W. M. Schmidt, Diophantine Approximations and Diophantine Equations. VIII, 217 pages. 1991.

Vol. 1468: J. Noguchi, T. Ohsawa (Eds.), Prospects in Complex Geometry. Proceedings, 1989. VII, 421 pages. 1991.

Vol. 1469: J. Lindenstrauss, V. D. Milman (Eds.), Geometric Aspects of Functional Analysis. Seminar 1989-90. XI, 191 pages. 1991.

Vol. 1470: E. Odell, H. Rosenthal (Eds.), Functional Analysis. Proceedings, 1987-89. VII, 199 pages. 1991.

Vol. 1471: A. A. Panchishkin, Non-Archimedean L-Functions of Siegel and Hilbert Modular Forms. VII, 157 pages. 1991.

Vol. 1472: T. T. Nielsen, Bose Algebras: The Complex and Real Wave Representations. V, 132 pages. 1991.

Vol. 1473: Y. Hino, S. Murakami, T. Naito, Functional Differential Equations with Infinite Delay. X, 317 pages. 1991.

Vol. 1474: S. Jackowski, B. Oliver, K. Pawałowski (Eds.), Algebraic Topology, Poznań 1989. Proceedings. VIII, 397 pages. 1991.

Vol. 1475: S. Busenberg, M. Martelli (Eds.), Delay Differential Equations and Dynamical Systems. Proceedings, 1990. VIII, 249 pages. 1991.

Vol. 1476: M. Bekkali, Topics in Set Theory. VII, 120 pages. 1991.

Vol. 1477: R. Jajte, Strong Limit Theorems in Noncommutative L_2-Spaces. X, 113 pages. 1991.

Vol. 1478: M.-P. Malliavin (Ed.), Topics in Invariant Theory. Seminar 1989-1990. VI, 272 pages. 1991.

Vol. 1479: S. Bloch, I. Dolgachev, W. Fulton (Eds.), Algebraic Geometry. Proceedings, 1989. VII, 300 pages. 1991.

Vol. 1480: F. Dumortier, R. Roussarie, J. Sotomayor, H. Żoładek, Bifurcations of Planar Vector Fields: Nilpotent Singularities and Abelian Integrals. VIII, 226 pages. 1991.

Vol. 1481: D. Ferus, U. Pinkall, U. Simon, B. Wegner (Eds.), Global Differential Geometry and Global Analysis. Proceedings, 1991. VIII, 283 pages. 1991.

Vol. 1482: J. Chabrowski, The Dirichlet Problem with L^2-Boundary Data for Elliptic Linear Equations. VI, 173 pages. 1991.

Vol. 1483: E. Reithmeier, Periodic Solutions of Nonlinear Dynamical Systems. VI, 171 pages. 1991.

Vol. 1484: H. Delfs, Homology of Locally Semialgebraic Spaces. IX, 136 pages. 1991.

Vol. 1485: J. Azéma, P. A. Meyer, M. Yor (Eds.), Séminaire de Probabilités XXV. VIII, 440 pages. 1991.

Vol. 1486: L. Arnold, H. Crauel, J.-P. Eckmann (Eds.), Lyapunov Exponents. Proceedings, 1990. VIII, 365 pages. 1991.

Vol. 1487: E. Freitag, Singular Modular Forms and Theta Relations. VI, 172 pages. 1991.

Vol. 1488: A. Carboni, M. C. Pedicchio, G. Rosolini (Eds.), Category Theory. Proceedings, 1990. VII, 494 pages. 1991.

Vol. 1489: A. Mielke, Hamiltonian and Lagrangian Flows on Center Manifolds. X, 140 pages. 1991.

Vol. 1490: K. Metsch, Linear Spaces with Few Lines. XIII, 196 pages. 1991.

Vol. 1491: E. Lluis-Puebla, J.-L. Loday, H. Gillet, C. Soulé, V. Snaith, Higher Algebraic K-Theory: an overview. IX, 164 pages. 1992.

Vol. 1492: K. R. Wicks, Fractals and Hyperspaces. VIII, 168 pages. 1991.

Vol. 1493: E. Benoît (Ed.), Dynamic Bifurcations. Proceedings, Luminy 1990. VII, 219 pages. 1991.

Vol. 1494: M.-T. Cheng, X.-W. Zhou, D.-G. Deng (Eds.), Harmonic Analysis. Proceedings, 1988. IX, 226 pages. 1991.

Vol. 1495: J. M. Bony, G. Grubb, L. Hörmander, H. Komatsu, J. Sjöstrand, Microlocal Analysis and Applications. Montecatini Terme, 1989. Editors: L. Cattabriga, L. Rodino. VII, 349 pages. 1991.

Vol. 1496: C. Foias, B. Francis, J. W. Helton, H. Kwakernaak, J. B. Pearson, H_∞-Control Theory. Como, 1990. Editors: E. Mosca, L. Pandolfi. VII, 336 pages. 1991.

Vol. 1497: G. T. Herman, A. K. Louis, F. Natterer (Eds.), Mathematical Methods in Tomography. Proceedings 1990. X, 268 pages. 1991.

Vol. 1498: R. Lang, Spectral Theory of Random Schrödinger Operators. X, 125 pages. 1991.

Vol. 1499: K. Taira, Boundary Value Problems and Markov Processes. IX, 132 pages. 1991.

Vol. 1500: J.-P. Serre, Lie Algebras and Lie Groups. VII, 168 pages. 1992.

Vol. 1501: A. De Masi, E. Presutti, Mathematical Methods for Hydrodynamic Limits. IX, 196 pages. 1991.

Vol. 1502: C. Simpson, Asymptotic Behavior of Monodromy. V, 139 pages. 1991.

Vol. 1503: S. Shokranian, The Selberg-Arthur Trace Formula (Lectures by J. Arthur). VII, 97 pages. 1991.

Vol. 1504: J. Cheeger, M. Gromov, C. Okonek, P. Pansu, Geometric Topology: Recent Developments. Editors: P. de Bartolomeis, F. Tricerri. VII, 197 pages. 1991.

Vol. 1505: K. Kajitani, T. Nishitani, The Hyperbolic Cauchy Problem. VII, 168 pages. 1991.

Vol. 1506: A. Buium, Differential Algebraic Groups of Finite Dimension. XV, 145 pages. 1992.

Vol. 1507: K. Hulek, T. Peternell, M. Schneider, F.-O. Schreyer (Eds.), Complex Algebraic Varieties. Proceedings, 1990. VII, 179 pages. 1992.

Vol. 1508: M. Vuorinen (Ed.), Quasiconformal Space Mappings. A Collection of Surveys 1960-1990. IX, 148 pages. 1992.

Vol. 1509: J. Aguadé, M. Castellet, F. R. Cohen (Eds.), Algebraic Topology - Homotopy and Group Cohomology. Proceedings, 1990. X, 330 pages. 1992.

Vol. 1510: P. P. Kulish (Ed.), Quantum Groups. Proceedings, 1990. XII, 398 pages. 1992.

Vol. 1511: B. S. Yadav, D. Singh (Eds.), Functional Analysis and Operator Theory. Proceedings, 1990. VIII, 223 pages. 1992.

Vol. 1512: L. M. Adleman, M.-D. A. Huang, Primality Testing and Abelian Varieties Over Finite Fields. VII, 142 pages. 1992.

Vol. 1513: L. S. Block, W. A. Coppel, Dynamics in One Dimension. VIII, 249 pages. 1992.

Vol. 1514: U. Krengel, K. Richter, V. Warstat (Eds.), Ergodic Theory and Related Topics III, Proceedings, 1990. VIII, 236 pages. 1992.

Vol. 1515: E. Ballico, F. Catanese, C. Ciliberto (Eds.), Classification of Irregular Varieties. Proceedings, 1990. VII, 149 pages. 1992.

Vol. 1516: R. A. Lorentz, Multivariate Birkhoff Interpolation. IX, 192 pages. 1992.

Vol. 1517: K. Keimel, W. Roth, Ordered Cones and Approximation. VI, 134 pages. 1992.

Vol. 1518: H. Stichtenoth, M. A. Tsfasman (Eds.), Coding Theory and Algebraic Geometry. Proceedings, 1991. VIII, 223 pages. 1992.

Vol. 1519: M. W. Short, The Primitive Soluble Permutation Groups of Degree less than 256. IX, 145 pages. 1992.

Vol. 1520: Yu. G. Borisovich, Yu. E. Gliklikh (Eds.), Global Analysis – Studies and Applications V. VII, 284 pages. 1992.

Vol. 1521: S. Busenberg, B. Forte, H. K. Kuiken, Mathematical Modelling of Industrial Process. Bari, 1990. Editors: V. Capasso, A. Fasano. VII, 162 pages. 1992.

Vol. 1522: J.-M. Delort, F. B. I. Transformation. VII, 101 pages. 1992.

Vol. 1523: W. Xue, Rings with Morita Duality. X, 168 pages. 1992.

Vol. 1524: M. Coste, L. Mahé, M.-F. Roy (Eds.), Real Algebraic Geometry. Proceedings, 1991. VIII, 418 pages. 1992.

Vol. 1525: C. Casacuberta, M. Castellet (Eds.), Mathematical Research Today and Tomorrow. VII, 112 pages. 1992.

Vol. 1526: J. Azéma, P. A. Meyer, M. Yor (Eds.), Séminaire de Probabilités XXVI. X, 633 pages. 1992.

Vol. 1527: M. I. Freidlin, J.-F. Le Gall, Ecole d'Eté de Probabilités de Saint-Flour XX – 1990. Editor: P. L. Hennequin. VIII, 244 pages. 1992.

Vol. 1528: G. Isac, Complementarity Problems. VI, 297 pages. 1992.

Vol. 1529: J. van Neerven, The Adjoint of a Semigroup of Linear Operators. X, 195 pages. 1992.

Vol. 1530: J. G. Heywood, K. Masuda, R. Rautmann, S. A. Solonnikov (Eds.), The Navier-Stokes Equations II – Theory and Numerical Methods. IX, 322 pages. 1992.

Vol. 1531: M. Stoer, Design of Survivable Networks. IV, 206 pages. 1992.

Vol. 1532: J. F. Colombeau, Multiplication of Distributions. X, 184 pages. 1992.

Vol. 1533: P. Jipsen, H. Rose, Varieties of Lattices. X, 162 pages. 1992.

Vol. 1534: C. Greither, Cyclic Galois Extensions of Commutative Rings. X, 145 pages. 1992.

Vol. 1535: A. B. Evans, Orthomorphism Graphs of Groups. VIII, 114 pages. 1992.

Vol. 1536: M. K. Kwong, A. Zettl, Norm Inequalities for Derivatives and Differences. VII, 150 pages. 1992.

Vol. 1537: P. Fitzpatrick, M. Martelli, J. Mawhin, R. Nussbaum, Topological Methods for Ordinary Differential Equations. Montecatini Terme, 1991. Editors: M. Furi, P. Zecca. VII, 218 pages. 1993.

Vol. 1538: P.-A. Meyer, Quantum Probability for Probabilists. X, 287 pages. 1993.

Vol. 1539: M. Coornaert, A. Papadopoulos, Symbolic Dynamics and Hyperbolic Groups. VIII, 138 pages. 1993.

Vol. 1540: H. Komatsu (Ed.), Functional Analysis and Related Topics, 1991. Proceedings. XXI, 413 pages. 1993.

Vol. 1541: D. A. Dawson, B. Maisonneuve, J. Spencer, Ecole d' Eté de Probabilités de Saint-Flour XXI - 1991. Editor: P. L. Hennequin. VIII, 356 pages. 1993.

Vol. 1542: J.Fröhlich, Th.Kerler, Quantum Groups, Quantum Categories and Quantum Field Theory. VII, 431 pages. 1993.

Vol. 1543: A. L. Dontchev, T. Zolezzi, Well-Posed Optimization Problems. XII, 421 pages. 1993.

Vol. 1544: M.Schürmann, White Noise on Bialgebras. VII, 146 pages. 1993.

Vol. 1545: J. Morgan, K. O'Grady, Differential Topology of Complex Surfaces. VIII, 224 pages. 1993.

Vol. 1546: V. V. Kalashnikov, V. M. Zolotarev (Eds.), Stability Problems for Stochastic Models. Proceedings, 1991. VIII, 229 pages. 1993.

Vol. 1547: P. Harmand, D. Werner, W. Werner, M-ideals in Banach Spaces and Banach Algebras. VIII, 387 pages. 1993.

Vol. 1548: T. Urabe, Dynkin Graphs and Quadrilateral Singularities. VI, 233 pages. 1993.

Vol. 1549: G. Vainikko, Multidimensional Weakly Singular Integral Equations. XI, 159 pages. 1993.

Vol. 1550: A. A. Gonchar, E. B. Saff (Eds.), Methods of Approximation Theory in Complex Analysis and Mathematical Physics IV, 222 pages, 1993.

Vol. 1551: L. Arkeryd, P. L. Lions, P.A. Markowich, S.R. S. Varadhan. Nonequilibrium Problems in Many-Particle Systems. Montecatini, 1992. Editors: C. Cercignani, M. Pulvirenti. VII, 158 pages 1993.

Vol. 1552: J. Hilgert, K.-H. Neeb, Lie Semigroups and their Applications. XII, 315 pages. 1993.

Vol. 1553: J.-L- Colliot-Thélène, J. Kato, P. Vojta. Arithmetic Algebraic Geometry. Trento, 1991. Editor: E. Ballico. VII, 223 pages. 1993.

Vol. 1554: A. K. Lenstra, H. W. Lenstra, Jr. (Eds.), The Development of the Number Field Sieve. VIII, 131 pages. 1993.

Vol. 1555: O. Liess, Conical Refraction and Higher Microlocalization. X, 389 pages. 1993.

Vol. 1556: S. B. Kuksin, Nearly Integrable Infinite-Dimensional Hamiltonian Systems. XXVII, 101 pages. 1993.

Vol. 1557: J. Azéma, P. A. Meyer, M. Yor (Eds.), Séminaire de Probabilités XXVII. VI, 327 pages. 1993.

Vol. 1558: T. J. Bridges, J. E. Furter, Singularity Theory and Equivariant Symplectic Maps. VI, 226 pages. 1993.

Vol. 1559: V. G. Sprindžuk, Classical Diophantine Equations. XII, 228 pages. 1993.

Vol. 1560: T. Bartsch, Topological Methods for Variational Problems with Symmetries. X, 152 pages. 1993.

Vol. 1561: I. S. Molchanov, Limit Theorems for Unions of Random Closed Sets. X, 157 pages. 1993.

Vol. 1562: G. Harder, Eisensteinkohomologie und die Konstruktion gemischter Motive. XX, 184 pages. 1993.

Vol. 1563: E. Fabes, M. Fukushima, L. Gross, C. Kenig, M. Röckner, D. W. Stroock, Dirichlet Forms. Varenna, 1992. Editors: G. Dell'Antonio, U. Mosco. VII, 245 pages. 1993.

Vol. 1564: J. Jorgenson, S. Lang, Basic Analysis of Regularized Series and Products. IX, 122 pages. 1993.

Vol. 1565: L. Boutet de Monvel, C. De Concini, C. Procesi, P. Schapira, M. Vergne. D-modules, Representation Theory, and Quantum Groups. Venezia, 1992. Editors: G. Zampieri, A. D'Agnolo. VII, 217 pages. 1993.

Vol. 1566: B. Edixhoven, J.-H. Evertse (Eds.), Diophantine Approximation and Abelian Varieties. XIII, 127 pages. 1993.

Vol. 1567: R. L. Dobrushin, S. Kusuoka, Statistical Mechanics and Fractals. VII, 98 pages. 1993.

Vol. 1568: F. Weisz, Martingale Hardy Spaces and their Application in Fourier Analysis. VIII, 217 pages. 1994.

Vol. 1569: V. Totik, Weighted Approximation with Varying Weight. VI, 117 pages. 1994.

Vol. 1570: R. deLaubenfels, Existence Families, Functional Calculi and Evolution Equations. XV, 234 pages. 1994.

Vol. 1571: S. Yu. Pilyugin, The Space of Dynamical Systems with the C^0-Topology. X, 188 pages. 1994.

Vol. 1572: L. Göttsche, Hilbert Schemes of Zero-Dimensional Subschemes of Smooth Varieties. IX, 196 pages. 1994.

Vol. 1573: V. P. Havin, N. K. Nikolski (Eds.), Linear and Complex Analysis – Problem Book 3 – Part I. XXII, 489 pages. 1994.

Vol. 1574: V. P. Havin, N. K. Nikolski (Eds.), Linear and Complex Analysis – Problem Book 3 – Part II. XXII, 507 pages. 1994.

Vol. 1575: M. Mitrea, Clifford Wavelets, Singular Integrals, and Hardy Spaces. XI, 116 pages. 1994.

Vol. 1576: K. Kitahara, Spaces of Approximating Functions with Haar-Like Conditions. X, 110 pages. 1994.

Vol. 1577: N. Obata, White Noise Calculus and Fock Space. X, 183 pages. 1994.

Vol. 1578: J. Bernstein, V. Lunts, Equivariant Sheaves and Functors. V, 139 pages. 1994.

Vol. 1579: N. Kazamaki, Continuous Exponential Martingales and *BMO*. VII, 91 pages. 1994.

Vol. 1580: M. Milman, Extrapolation and Optimal Decompositions with Applications to Analysis. XI, 161 pages. 1994.

Vol. 1581: D. Bakry, R. D. Gill, S. A. Molchanov, Lectures on Probability Theory. Editor: P. Bernard. VIII, 420 pages. 1994.

Vol. 1582: W. Balser, From Divergent Power Series to Analytic Functions. X, 108 pages. 1994.

Vol. 1583: J. Azéma, P. A. Meyer, M. Yor (Eds.), Séminaire de Probabilités XXVIII. VI, 334 pages. 1994.

Vol. 1584: M. Brokate, N. Kenmochi, I. Müller, J. F. Rodriguez, C. Verdi, Phase Transitions and Hysteresis. Montecatini Terme, 1993. Editor: A. Visintin. VII. 291 pages. 1994.

Vol. 1585: G. Frey (Ed.), On Artin's Conjecture for Odd 2-dimensional Representations. VIII, 148 pages. 1994.

Vol. 1586: R. Nillsen, Difference Spaces and Invariant Linear Forms. XII, 186 pages. 1994.

Vol. 1587: N. Xi, Representations of Affine Hecke Algebras. VIII, 137 pages. 1994.

Vol. 1588: C. Scheiderer, Real and Étale Cohomology. XXIV, 273 pages. 1994.

Vol. 1589: J. Bellissard, M. Degli Esposti, G. Forni, S. Graffi, S. Isola, J. N. Mather, Transition to Chaos in Classical and Quantum Mechanics. Montecatini, 1991. Editor: S. Graffi. VII, 192 pages. 1994.

Vol. 1590: P. M. Soardi, Potential Theory on Infinite Networks. VIII, 187 pages. 1994.

Vol. 1591: M. Abate, G. Patrizio, Finsler Metrics – A Global Approach. IX, 180 pages. 1994.

Vol. 1592: K. W. Breitung, Asymptotic Approximations for Probability Integrals. IX, 146 pages. 1994.